Of Natural and Supernatural Things, and Others

Basilius Valentinus, Roger Bacon and John Isaac Holland

Translated by Daniel Cable

Esprios.com

BASILIUS VALENTINUS,

A

BENEDICTINE MONK,

OF

NATURAL & SUPERNATURAL

THINGS.

ALSO,

Of the first *Tincture*, *Root*, and *Spirit* of METALS and MINERALS, how the same are *Conceived*, *Generated*, *Brought forth*, *Changed*, and *Augmented*. Frier Roger Bacon, of the Medicine or Tincture of Antimony; Mr. John Isaac Holland, his Work of Saturn

Translated out of *High Dutch* by
DANIEL CABLE.

1671.

Of Natural and Supernatural Things, and Others

BASILIUS VALENTINUS,

OF

NATURAL AND SUPERNATURAL THINGS.

CHAP. I.

Because I have at this present undertaken to write of the of the first Tincture, the Root of Metals and Minerals, and to inform you of the Spiritual Essence, how the Metals and Minerals are at first spiritually conceived and born corporally; it will be necessary first of all to utter, and to acquaint you by a speech, that all things consist of two parts, that is, Natural and Supernatural; what is visible, tangible, and hath form or shape, that is natural; but what is intactible, without form, and spiritual, that is supernatural, and must be apprehended and conceived by Faith; such is the Creation, and especially the Eternity of God without end, immensible and incomprehensible; for Nature cannot conceive nor apprehend it by its humane reason: This is supernatural, what Reason cannot apprehend, but must be conceived by Faith, this is a Divine matter, and belongs to Theology, which judgeth Souls. Moreover, there appertains to supernatural things, the Angels of the Lord, having clarified Bodies, doing that by the permission of their Creator, which is impossible for any other Creature to do, their Works being concealed from the Eyes of the World, and so likewise are the Works of the Infernal Spirits and Devils unknown, which they do by the permission of the most High God. But above all the great Works of God are found and acknowledged to be supernatural, not to be scann'd and comprehended by Humane Imaginations; such is in especial the great Grace and Mercy of God which he bestows upon Mankind out of his great Love, which indeed no man can apprehend or know, and other great and wonderful works which he hath manifested divers manner of wayes by Christ our Saviour and Redeemer, for the confirmation of his Omnipotence and Glory: As when he raised *Lazarus* from the dead, *Jairus* his Daughter, the Ruler of the Synagogue, and the Widows Son of *Naim*. He made the Dumb to speak, the Deaf to hear, and the Blind to see, all which are supernatural, and *Magnalia Dei*; so also was his Conception, Resurrection, Descension, and Ascension into Heaven, too deep and mysterious for Nature; all which is only to be obtained by Faith.

There belongs likewise to supernatural things, the taking of *Enoch* and *Elias* into Heaven, the divine rapture of St. *Paul* in the Spirit into the third Heaven. Moreover, many supernatural things are done by Imagination, Dreams, and Visions; many wonders are done by the Imagination, witness the speckled Sheep by the speckled Rods laid in their watring places. God warned the wise men of the East by an especial Dream not to return again to *Herod*; likewise their three Persons, their three Gifts, Presents, or Offerings, and the supernatural Star, have all their peculiar and mystical meaning.

Nor was that Dream which hapned to *Pilates* Wife natural, who unjustly adjudged our Lord and Saviour Jesus Christ to death. The Vision of the Angels which appeared to the Shepherds at the Birth of Christ, and to the Women at his Sepulchre, who sought his Body where they had laid it, cannot be accounted Natural.

There are many other supernatural things done at several times by the Prophets & Saints; so was the voice of the Ass speaking to *Balaam*, contrary to the common course of Nature; as also *Joseph's* Interpretation of Dreams. And so God by his Angels preserves us oftentimes from infinite Evils, and delivers us out of manifold Dangers, impossible for Nature to do.

All this & many others belong to Theology, and to Heaven, whereunto the Soul is to have regard. Now follows the supernatural things of the visible Works of God, as we see them in the Firmament; to wit, the Planets, Stars, and Elements, which are above our Reason, only their Course and Motion is observed by speculation and reckoning, which belongs to Astronomy; it is a visible but incomprehensible Being, performing its operation in a Magnetick way, out of which likewise divers admirable things are found and observed, which are altogether supernatural; understand it thus, that the Heaven operates in the Earth, and the Earth affords a correspondence with the Heavenly. For the Earth hath also its seven Planets, which are operated and bred by the seven Celestial, only by a spiritual Impression or Infusion, even as the Stars operate all

Of Natural and Supernatural Things, and Others

Minerals. This is done incomprehensibly and spiritually, and therefore it is to be accounted supernatural, even as two Lovers, their persons are visible, but their Love one to the other is invisible: Humane Bodies are tangible and natural, but Love is invisible, spiritual, intangible and supernatural, comparable to a Magnetick Attraction only; for the invisible Love which is attracted unto it spiritually by the Imagination is, accomplish'd by the desires and fruition. In like manner when the Heaven hath a love to the Earth, and the Earth hath a Love, Inclination, and Affection towards Man, as the great World to the lesser, for the lesser World is taken out of the greater, and when the Earth by the desires of its invisible Imagination doth attract unto itself such a Love of the Heaven, there is thereby an Union of the Superiour and Inferiour, as Man and Wife are accounted one Body together, and after this Union the Earth is impregnated by the Infusion of the Heaven, and begins to conceive and bring forth a Birth sutable to the Infusion, and this Birth after its Conception is digested by the Elements, and brought to a perfect Ripeness and this is reckoned among the supernatural things; how the supernatural Essence performs its operation in the natural.

Among the supernatural things are likewise reckoned all Magical and Cabalistical Matters which depend thereon, arising out of the Light of true knowledge, not those which proceed from Superstition, Conjuration, or unlawful Exorcisme, such as the Sorcerers use; but I mean in this place such a Magick as the Wise men had that came out of the *East*, who by Revelation from God, and by true allowable Art judged rightly; or such an one, as those of old had before us, usual among the *Egyptians* and *Arabians*, before Writing was found, they noted, observed, and reserved by Signs, Characters and Hieroglyphicks. Such Blessings may be used, which Christ the Son of God used, as the Scripture saith; He took little Children, laid his hands upon them, and blessed them. But whatsoever is contrary to God and his Word, ought justly to be rejected, and not to be tollerated, because they are not Godly, but Diabolical. But those Supernatural things which oppose not God and his Holy Word, belong unto Magick, and do the Soul no prejudice.

As concerning Visions which Holy Men of God have often seen, it is reason they should be reckoned among those things which are not

Of Natural and Supernatural Things, and Others

Natural; for whatsoever man speculates and comprehends by the Mind, is Supernatural; on the contrary, whatsoever he can take, see, and hold is Natural.

Let us consider the third part of Natural & Supernatural things in Physick, the Virtues and Powers of each; this Medicine of every thing must first be driven out of a visible, tangible, natural Body, and be brought into a spiritual, meliorated, supernatural operation, that the Spirit which at the first was infused and given to the Body to live, might be released, that it should operate and penetrate as a Spiritual Essence, and Fire, having its Vent-holes left that it might burn and have no opposition, which might suffocate, suppress, or hinder the burning Life; whereas otherwise, if the separation of the Soul and Spirit from the Body should not first be done, there could not succeed any operation either effectual, profitable, or necessary; for whatsoever is visible, to be felt, and inseparably in a Body, that is Natural and Corporal; but so soon as there is a separation, the living departs from the dead, gaining its perfect operation, and the natural Body being separated, the spiritual Essence is free to penetrate, becoming a spiritual and supernatural Medicine. In brief, all things (none excepted) which we can touch and handle, are natural, but they must be made supernatural, if you would prepare them for Physick; for the supernatural only hath a living power in it to operate, the natural hath only a dead tangible Form. For when *Adam* was made, he was dead, having no life of any virtue, but so soon as the operative quickning Spirit came to him, then he manifested his living virtue and power by supernatural admiration, so that in every thing there is the natural and supernatural united in one, and bound together in their habitation, that every thing might be perfect; for all created things in the world are some supernatural, only what concerns the Soul and spiritual matters, and some are natural and supernatural, with what concerns the Elements and Firmament, as likewise the Minerals, Vegetables, and Animals, which is known and found, when they are separated one from the other, that the Soul departs out of the Body, and the Spirit forsakes its Soul, leaving the Body an empty habitation.

Moreover, you must understand and consider, that the great and little World are made and formed of one first Matter, by an

Of Natural and Supernatural Things, and Others

unsearchable Almighty Essence, at that time in the beginning, when the Spirit of God moved upon the Water, who was from Eternity without beginning. The great World, as Heaven and Earth, was first, then was Man, the Little World, taken out of the greater; the Water was separated from the Earth, the Water was the Matter whereon the everlasting Spirit of God moved; the Little World was formed of the noblest Earth, as its Quintessence, by the Aquosity which yet was in the Earth, and all was only Natural; but after the breathing in of the Divine heating Breath, immediately the Supernatural was added; so then the Natural and Supernatural were knit and united. The great World is perishable, yet there will be a New Earth or World; the Little World is Eternal, the Great, Created, Dissoluble world will again be brought to nothing, but the little world will be clarified by the Spirit of God, because he possesses it, making a Celestial clarified water out of the aforesaid Earthly water; then it will follow, that the first matter will be turned into the last, and the last matter will become the first. Now the reason why the great world is perishable, is this, that the Spirit of God hath not his dwelling or habitation in the great world, but in the little world; for Man is the Temple of the Holy Ghost, if he do not wilfully defile himself, adhering to the Hellish Fire, which makes a breach and difference. For he remaines in the little world, which he formed after his own similitude, and made him a consecrated Temple; otherwise there is every thing in the little world which is to be found in the great, as Heaven and Earth with the Elements, and what depends thereon, or appertains thereunto.

We find also that in the first Creation, which was performed of nothing, three things arose; to wit, a Soulish, Spiritual, Invisible Essence, which represented a Mercurial Water, a Sulphurous Vapour, and a Terrene Salt; these three gave a compleat and perfect, a tangible and formal Body to all things wherein especially all the four Elements are contained, as I have already mentioned in my Writing where I treat of the Microcosme.

But that I may yet give a little more information of Natural and Supernatural things, as well spiritual as corporal: We find that the *Canaanitish* Woman was cured of her Flux of Blood which held her twelve years, only by a bare touch, when she touched the Garment of the Son of God, her Disease being natural, but the Medicine or Cure was Supernatural, because by her Faith she gained help from the Lord Christ.

Likewise we have an excellent, high and supernatural Miracle in the three Children, *Shadrach, Meshach,* and *Abednego,* who were cast into the fiery Furnace, by the Command of King *Nebuchadnezzar,* yet by God wonderfully delivered, and not consumed, *Dan.* 3.

So also the Confusion of Tongues, and Infusion of divers Speeches at the foolish structure of the Tower of *Babel,* which should have reached up to Heaven, is esteemed for a Supernatural Miracle. And so was that a Supernatural Sign, when the Children of *Israel* did lap water as Dogs do, when a small Number at Gods Command, fought against the *Midianites,* Judg. 7. 6. So the sending of the Dove by *Noah* out of the Ark, when she brought an Olive Branch in her Bill, a Sign of Mercy, and a Divine Supernatural Message.

When the Holy man of God *Moses* struck the Rock with his Rod, that the hard Rock yielded Water, is beyond humane Reason; so was the turning of the salt water into sweet and drinkable, supernatural. As also the dry passage of the Children of *Israel* through the Red Sea; and the Budding of *Aarons* Rod, are all supernatural. In brief, the Resurrection of Christ the Eternal Son of God out of the grave, for all the Tomb-stone, his appearing to the two men going to *Emas,* his revealing himself to his Disciples when the Door was lock'd, are all Divine and Supernatural. Divers Examples more might be recited out of Divine Writ, which for Brevities sake I omit.

Among Supernatural things are accounted all Mineral Signs, as the Appearance of Spirits, Representations, Pigmies appearing diversly and numerously, giving notice of good or bad Luck, Ruine or Riches; so also those Figures, Shapes, or other works found in the Ores of Metals, as of Men, Fishes, and other Creatures, so formed and represented by the imagination of the three first Principles, then ripened and fully digested by the Earth, and other Elements. Hereunto appertain the Monsters of the Earth, and such things as are found within the Earth at certain times of a wonderful form and shape, but not at all to be found when that time is past, yet appear again and are to be found at some other time.

Of Natural and Supernatural Things, and Others

Hereunto also belong all Visions and Appearances performed by Water, Glasses, Cristal, or other means, as also those done by Sigils and Characters, which yet are so various, some being only Natural, yet affording Supernatural appearances or sights; but the others which are performed by Conjurations, are neither Natural nor Supernatural, but Diabolical, belonging unto Sorcery, and are prohibited all good Christians; so likewise all those Means which oppose Holy Writ, Gods Word and Commandments, are to be rejected and refuted by true Natural Cabalists; I say this, because a certain distinction and sure order ought to be found of the Natural, Supernatural, Unnatural things.

In like manner there appertains unto Supernatural things, all the Water-Spirits, as the *Syrens*, *Succubi*, & other Water-Nymphs, with their Relations, as likewise the Terrestrial Spirits, and those which inhabit the Air, who sometimes are heard, seen, or perceived, sometimes foretelling Death or other Disasters, sometimes they discover by their Apparition Riches and good Fortune in certain places, and the Fiery Spirits appertain here also, which appear in a fiery shape, or like a burning Light; all these are Spirits having untangible Bodies, yet are they not such Spirits as the right Hellish Spirits, who hunt after mens Souls as an Eternal Jewel, even as the Infernal *Lucifer*, the Devil and his Dependents do, who were ejected with him; but these are such Spirits which are above Nature, set before Men for admiration, and are only maintained by the Elements, whereby they are nourished and fed; but when this Earthly world shall cease, they also shall decay and vanish with it, because they have no Souls to be saved. I will say no more hereof at present, but refer the opening of such Circumstances more at large to another opportunity, where I shall particularly treat of Visions and Spiritual Appearances, which are esteemed Unnatural by most part of the World, yet truly are Natural, but they are found to be Supernatural in their Operations and wonderful Qualities.

That I may further confirm my Assertion, I say likewise, that there are many things to be found in Physick, which yield and manifest their workings supernaturally in a Magnetical way, operating only by an attractive spiritual power which is attracted to it by the Air; for the Air is the *Medium* between the Physick and the Hurt or

Distemper, even as the *Magnet* ever doth direct and turn it self towards its Polestar, though the star be many thousand Miles distant from it, yet the spiritual operation and sympathy between them is so prevalent, that it is attracted together at so vast a distance by the *Medium* or middle Band of the Air; but because this attractive power is only known unto people in general, or as a thing common, it is therefore become customary, and is so esteemed, there being no notice taken of any further Secret whence this operative Faculty hath its Source or Original: In like manner Hurts and Distempers may be healed and cured, though the Patient and Physitian be very far distant one from the other; not by Charmes, Exorcismes, or other unlawful prohibited means, which are opposite to God and Nature, but by such means wherein the attractive Magnetick Virtue lies to accomplish it. As when a wounded person goes a Journey, leaving the Weapon wherewith he was wounded, or else of his Bloud which issued out of the wound with his Physician, wherewith he proceeds rightly and by orderly means, as is usual in dressing a wound, without all doubt he shall be absolutely cured, this is no Witchcraft, but the cure is performed only by the attractive power of the Medicine, which is carried to the Sore by the means of the Air, wherewith it is mundified, that it may perform the Spiritual Operation.

Some will think these hard sayings, and impossible in Nature, and many will say it is contrary to Nature, whereby many will be excited to dispute it, and raise Arguments one opposite to the other, whether it be Natural or no, whether it be possible or no, or whether it be Sorcery: I will thus resolve them, that this Cure is natural, but as it operates it is supernatural & spiritual, because it is performed meerly by an attractive incomprehensible means, and that this manner of Cure is no Sorcery: I affirm it hereby, that it is not mixt or accompanied with any Sorcery, nor with any other unnatural Means, contrary to God the Creator, or his holy and saving Word. But it is only Natural, out of its supernatural, invisible, incomprehensible, spiritual, and attractive power, which received its Original from the Sydereal, and performs its Operation by the Elements.

Lastly, I likewise approve this Cure to be no Sorcery, because the Devil rather delights in all Mischief to Mankind, than to assist any

Of Natural and Supernatural Things, and Others

manner of way for their benefit, which yet is impossible for him to do without Gods permission. Much more might be written of this Magnetick Form, but I chuse rather to be silent; referring it till I come to treat of the Natural Miracles of the World.

The grosser sort of foolish Wits, who imagine themselves to be wise Philosophers, and all others who are not in their perfect senses, know no difference in this case, but the wise and truly discreet well know how to distinguish betwixt that which is natural and that which is supernatural.

For do but observe this comparison, to be proved by a gross Example, how many Creatures are there which dye absolutely in the water, so that no life is left therein, but so soon as the pleasant Summer appears, the natural heat gives a new life, & the Body quite restored in the same substance as it was before in its living Motion; even as an Herb, which dies in the Winter, but in the Spring it manifests it self anew. The death of these things is to be esteemed natural, but the return of a new life in its knowledge is supernatural; but because we are accustomed to all these things, the least part of us consider what is worthy of further Meditation in this case, letting both natural and supernatural go away together.

Most people overpass, that natural custom which yet is supernatural, as also monstrous Births, and those that bring signs and marks with them into the world; which may all be natural, but manifest themselves supernaturally, by the imagination which caused them: These supernatural forms and customs, the Mother of the Child caused by intervening thoughts, which unexpectedly happened to her, as it were by accident: Even as we often see and find, that many Men naturally are born with some gestures, which he can never leave, though he endevour with all his might to do it. The natural gestures of these Men are natural, but the conception in the Womb which caused the imagination of these things is supernatural, and subject to what the Heaven imprints.

To conclude: I say, that none can defend the supernatural not to be true by good grounds and reasons, except he have learn'd to know the natural, which hath its original, and gained its shape from the supernatural; after he hath learned this, he may evidence it by sure proofs, that he will be conquerour over those, who will not believe

Of Natural and Supernatural Things, and Others

what is supernatural; and he will convince the opinions of those who dispute of natural things, and yet know not the grounds, saving only a bare pretence, much talk, tedious and unprofitable Debates.

Chap. II.

Of the first Tincture and Roots of Metals.

But now to come to my intent, and by Gods permission to accomplish the same. I undertake to certifie of the first Tincture, Root, and Generation of Metals and Minerals: Know that the first Tincture and Root of all Metals, is likewise a supernatural, flying, fiery Spirit; which preserves it self in the Air, seeking its habitation naturally in the Earth and Water, wherein it can rest and operate: This Spirit is found in all Metals, more abundant in other Metals than in Gold, because Gold, by reason of its well digested, ripened, and fixt body, is tight, close, and compact, and therefore no more can enter into its body than is just requisite; but the other Metals have not such fixt bodies, for their pores are open, and far extenuated, therefore the Tincture Spirit can the more abundantly pass thorough and possess them. But because the bodies of the other Metals are inconstant, the Tincture cannot remain with those inconstant bodies, but must depart. And whereas the Tincture of Gold is found in none more plentiful than in *Mars* and *Venus*, as Man and Wife, their bodies therefore are destroyed, and the tinging Spirit taken out of them, which makes Gold sanguin, being first opened and prepared, and by their food and drink it becomes volatile, wherefore this volatile Gold being satisfied with its food and drink, assumes its own bloud to it self, dries it up by its own internal heat, by the help and assistance of the vaporous fire, and there is a Conquest again, which is quite fix'd, makes the highest Constancy, that the Gold becomes an over-fix'd Medicine, by reason of abundance of Bloud it yields no Body, except another superfluous Body be again put to it, wherein the abounding fix'd bloud may disperse itself, this additional Metallick Body, by reason of the great heat of the fix'd Lions bloud, is penetrated as by fire, and purged from all impurity, and forthwith throughly digested to a perfect ripeness and fixedness: That first of all the Servant brings the Matter unto Riches, because the Master before could not spare any of his Cloaths to give away, seeing that Nature had lent and endowed him with one Noble Suit only; on the other side, the King, when he hath received his Aides and Contributions from his

Subjects, can then distribute possessions, and permanent Liveries, that the Lord and Servant may remain both together; and do not think it strange, that the King needs to borrow of his Subjects, because their Bodies are unfix'd and inconstant, for they receive much, and yet can keep but little Credit: But if the King can participate thereof, he will the better overcome heat and Frost, than the Leprous Metals can; and henceforth by this Receipt he becomes particularly a Dominator and Conquerour of all other, with a great Victory and triumph of Riches and of Health to long Life. I hope you have from the beginning sufficiently understood concerning this Natural and Supernatural Advice, and the first tinging Root of Metals and Minerals, whereon the Corner-stone is placed, and where the true Rock is grounded in its kind, wherein Nature hath placed and buried her secret & deeply concealed Gifts; to wit, in the fiery tinged Spirits, which Colours they gained out of the starry Heaven by the operation of the Elements; and they can moreover tinge and fix that which before was not tinged and unfix'd, seeing that *Luna* wants the Robe of the Golden Crown, together with the fixedness, as likewise *Saturn*, *Jupiter*, and *Mercury* do; and although *Mars* and *Venus* need not this Rayment, but can communicate it to the other five, yet I say, that they can perform nothing to attain any thing with wealth without the Lion, because they are not sufficiently accomodated with a fixedness of their *Mercury*, and a gentleness of their Salt, except it be that the Lion overcome them, that they have triumphed on both parts, and gained a remarkable Melioration altogether; this Melioration lies concealed in their Signate star, or Magnet, out of which all Metals have themselves received their Gifts.

Now I will proceed, and particularly step to the Birth and Generation, how the or *Archæus* manifests its power; pouring it forth, and daily reveales it, whereby all Metallick and Mineral Forms are visibly proposed, and made formal, tangible, and corporal by the Mineral, intangible, flying, fiery Spirits: Understand therefore further, and observe with diligence and care, that by forgetfulness you let not that which is weighty pass away, nor yet neglect or overlook that which is most profitable, and on the contrary observe the bare words at length, passing over the Truth; for what I write herein, is undoubtedly held and esteemed that the highest is undoubtedly by many esteemed for the lowest, and the lowest for the highest Mystery, and is so to be reputed.

Now you must first know, that all Metals and Minerals of the Earth have one only Matter and Mother, whereby in general they all received Conception, gaining a compleat and corporal Birth. This Matter which comes out of the Center, first of all divides it self into three parts, to procure one corporal or certain form of each Metal. These three parts are only fed in the Earth by the Elements, out of their Bodies, and nourish'd till they be perfect. But the Matter which comes out of the Center is imagined by the Stars, operated by the Elements, and formed by the Earth: It is a Matter to be known, and the true Mother of Metals and Minerals: It is such a Matter and Mother, whereof Man himself is conceived, born, nourish'd, and made corporal: It may be compared to the middle World, for what is in the great World is in the little World, and what is in the little World is also in the greater; and what is jointly in the great and little World is likewise found in the middle World, which unites and conjoins the great and little world; it is a Soul which unites and copulates the Spirit with the Body. This Soul is compared unto water, and it is a right true water, but not so that it wets as other water doth, but it is a Celestial water, dry, found in a Metallick Liquorish substance; it is a Soulish water, which loves all Spirits, and unites them with their Bodies, conducting them to a compleat Life; therefore it is reasonably found out, and evidently proved, that Water is the Mocker of all Metals, which are heated by the warm aerial Fire, or Spirit of *Sulphur*, which by its digestion makes the Earthly Body lively, wherein the Salt is evidently found, which preserves from putrefaction so that nothing might be consumed by Corruption. At the beginning and birth *Quick-silver* is first operated, which stands yet open with a subtile coagulation, because little Salt is imparted to it, whereby he manifests a more spiritual than corporal Body; but all the other Metals which follow out of its Essence, and have more Salt, whereby they become corporal, do all follow this; so that I now begin first with the Spirit of *Mercury*.

Chap. III.

Of the Spirit of Mercury.

Though I have a peculiar Stile in writing, which will seem strange unto many, causing strange Thoughts and Fancies in their Brains, yet there is reason enough for my so doing; I say enough, that I may remain by my own experience, not esteeming much of others prating, because it is concealed in my knowledge, Seeing having alwaies the preheminence before Hearing, and Reason hath the praise before Folly; Wherefore I now say, that all visible, tangible things are made of the Spirit of *Mercury*, which excels all earthly things of the whole world, all things being made out of it, having their Off-spring only from it; for all is found therein which can perform all whatsoever the Artist desires to find; It is the beginning to operate Metals, when it is become a spiritual Essence, which is meer Air flying to and fro without wings; it is a moving wind, which after it is expelled its dwelling by *Vulcan*, it is driven into its *Chaos*, where it again enters, and resolves it self into the Elements, where it is elevated and attracted by the Sydereal Stars after a Magnetical manner unto themselves, out of love, whence he proceeded before, and was operated, because it affects its like again, and attracts it to it. But if this Spirit of *Mercury* can be caught, and made corporal, it resolves into a Body, and becomes a pure, clear, transparent water, which is the true spiritual water, and the first *Mercurial* Root of the Minerals and Metals, spiritual, intangible, incombustible, without any mixture of earthly Aquosity; it is that Celestial water, whereof very much hath been written; for by this Spirit of *Mercury* all Metals, may if need require, be broken, opened, and resolved into their first Matter, without Corrosive; it renews the age of Man or Beast, even as the Eagles; it consumes all evil, and conducts a long Age to long Life. This Spirit of *Mercury* is the Master-Key of my Second Key, whereof I wrote in the beginning; wherefore I will call; *Come ye Blessed of the Lord, be anointed, and refreshed with water, and embalm your Bodies, that they may not putrefie or stink*; for this Celestial Water is the beginning, the Oyl, and the means, seeing it burns not, because it is made of a spiritual Sulphur, the Salt Balsam is corporal, which is united with the Water by the Oyl, whereof I will afterwards treat more at large, when I shall write of them, and mention them.

And that I may further declare what is the Essence, Matter and Form of the Spirit of *Mercury*, I say, that its Essence is blessed, its Matter spiritual and its form earthly, which yet must be understood by an incomprehensible way; these are indeed harsh Expressions, many will think, thy Proposals are all vain, strange Effusions, raising wonderful Imaginations, and true it is that they are strange, and require strange people to understand these Sayings; it is not written for Peasants, how they should grease Cart-wheels, nor is it written unto those who have no knowledge of the Art, though they be never so learned, or think themselves so; for I only account them Learned, who next unto Gods Word, learn to know Earthly things, which must be pondered and judged by the Understanding, founded upon a true Knowledge, to distinguish Light from Darkness, who chuse that which is good, and reject the evil.

It is needless for you to know what the beginning of this Spirit of *Mercury* requires, because it can in no wise help nor advantage you, only take notice of this, that its beginning is supernatural, out of the Celestial, Sydereal and Elementary, bestowed on it from the beginning of the first Creation, that it may enter further into an Earthly Substance. But because this is necessary which hath been declared to you, leave the Celestial to the Soul, apprehend it by Faith, and let the Sydereal likewise alone, because these Sydereal Impressions are invisible and intangible, the Elements have already brought forth the Spirit perfect into the world by the Nutriment, therefore let that alone likewise; for man cannot make the Elements, but only the Creator, and remain by thy made Spirit which is already formal and unformal, tangible and intangible, and yet is presented visibly. So have you enough of the first Matter, out of which all Metals and Minerals grow, and is one only thing, and such a matter which unites it self with the *Sulphur* in the following Chapter, and enters into a Coagulation with the *Salt* of the fifth Chapter, that it may be one Body, and a perfect Medicine of all Metals, not only to bring forth in the Earth at the beginning, as in the great World, but also by help of the vaporous Body to transmute and change, together with the augmentation in the lesser World: Let not this seem strange to you, seeing the Most High hath permitted, and Nature undertaken it.

Many will not believe this, esteeming it impossible, despise and vilifie these Mysteries, which they understand not in the least, they may remain Fools and Idiots till an illumination follows, which cannot be without Gods Will; but remains till the time predestinate. But wise and discreet, men who have truly shed the sweat of their Brows, will be my sufficient witnesses, and confirm the Truth, and indeed believe and hold for a truth all that which I write in this case, as true as Heaven and Hell are preordained, and proposed as Rewards of good and evil to the Elect and Reprobate. Now I write not only with my hands, but my Mind, Will and heart constrain me to it: Those who are highly conceited, illuminated, and world-wise, hate, envy, scandalize, defame and persecute this Mystery to the utmost Rind, or innermost Kernel, which hath its beginning out of the Center; but I know assuredly, there will come a time, when my Marrow is wasted, and my Bones dried up, that some will take my part heartily, after I am in the Pit; and if God would permit it, they would willingly raise me from the dead; but that cannot be; wherefore I have left them my Writings, that their Faith and Hope may have a Seal of Certainty and Truth, to testifie of me what my last Will and Testament was, which I ordained for the poor, and all the Lovers of Mysteries, though it did not behove me to have wrote so much, yet I could not refrain without prejudice to my Soul, but to drive a Light or Flash through a Cloud, that the Day might be observed, and the dark Night, thick and gloomy, rainy Weather expelled.

Now how the *Archæus* operates further by the Spirit of *Mercury* in the Earth, or Veins of the Earth, take this Advice, that after the spiritual Seed is formed by the impression of the Stars from above, and fed by the Elements, it is a Seed, and turns it self into a *Mercurial* Water, as first of all the great World was made of nothing, for when the Spirit moved upon the Water, the Celestial Heat must needs raise a Life in the cold watrish and earthly Creatures; in the great World it was Gods Power, and the Operation of the Celestial Lights; in the little World it is likewise Gods Power, and the Operation to work into the Earth by his Divine and Holy Breath. Moreover the

Almighty gave and Ordained means to accomplish it, that one Creature had obtained power to operate in the other, and the one to help and assist the other, to perform and fulfil all the Works of the Lord; and so an influence was permitted the Earth to bring forth by the Lights of Heaven, as also an internal Heat, to warm and digest that which was too cold for the Earth, by reason of its humidity, as unto every Creature a peculiar fashion according to its kind; so that a subtile sulphurous Vapour, is stired up by the Starry Heaven, not the common, but another more clarified and pure Vapour, distinct from others, which unites it self with the *Mercurial* Substance; by whose warm property, in process of time, the superfluous Moisture is dryed up, and then when the foulish property comes to it, which gives a preservation to the Body and Balsam, operating first into the Earth by a spiritual and sydereal influence, then are Metals generated of it, as it pleaseth the Mixture of the three Principles, the Body being formed according as it assumes unto it the greatest part of those three. But if the Spirit of *Mercury* be intended and qualified from above upon Animals, it becomes an Animal Substance; if it goes upon Vegetables by order, it becomes a Vegetable Work; but if, by reason of its infused nature, it fall on Minerals, it becomes Minerals and Metals, yet each one hath its distinction as they are wrought, the Animals for themselves, the Vegetables, on another manner and form by themselves, and so likewise the Minerals, each one a several way, whereof to write particularly would be too tedious, and yield large and various Narrations.

Many one may here demand and not without cause, how such a Spirit of *Mercury* may be procured, how to be made, and after what manner it is to be prepared to expel Diseases, and change all the kinds of the meaner and baser Metals, as if they were born in a little world, by transmutation and augmentation of their Seed; many expect this with impatience. I answer without concealing any thing, but will truly discover as much as is permitted me by Gods Command, in manner and form following.

In the Name of the Lord, Take a Red Quick-silver Ore which is like unto *Sinople* (or *Vermilion*) and the best Gold Ore you can get; grind of each a like quantity both together, before they partake of any fire, poure an Oyl of *Mercury*, upon it made *per se*, of common, purified and sublimed Quicksilver, set it a month to digest, you have an Extract rather Celestial than Terrestrial; distil this Extract gently, as in *Balneum Mariæ*, the Flegme ascends over, the Oyl remaining at bottom, being heavy, which in a moment receives all Metals into it poure thrice as much Spirit of Wine to it, circulate it in a Pellican, till it be as red as Bloud, and become so sweet that nothing may compare with it; decant the Spirit of Wine to a Liquidness, poure fresh Spirit of Wine upon it, this reiterate so often, till the Matter be exceeding sweet, and transparent red as a a Ruby, then put all together, poure that which ascended over upon white calcined *Tartar*, and distil it strongly in Ashes, the Spirit of Wine remains behind with the *Tartar*, but the Spirit of *Mercury* ascends over. If this Spirit of *Mercury* be mixt with the Spirit of *Sulphur*, together with its Salt, and so brought over jointly together, that they can never be separated, you have such a work which if it come over, and it get its ferment with Gold by solution according to a just measure and time appointed, and be brought to a perfect ripeness, unto the *Plusquam* perfection, nothing may compare therewith, for prevention of Diseases, and poverty, and to a rich excessive recreation of the Body and Goods. This is the way to obtain the Spirit of *Mercury*, which I have revealed as far as it is permitted me to do, by the Supremest Emperour; the Manual Operations are found in the Work which I have revealed; you must wisely observe, that you may not endure a Bath in Hell for me, by my true admonition to thee, forasmuch as a true opening of the Door which leads to the Royal Palace, is performed but with one Key, which cures all Diseases, be it *Dropsie, Consumption, Gout, Stone, Falling Sickness, Apoplexy, Leprosie,* or howsoever called in general: This Medicine likewise cures all kinds of the *French Pox*, and all old Sores of long standing, be it *Wolf, Noli me tangere, Tetter, Ring-worm, Cancer, Fistula,* and corroding hollow Sores; all which I have declared, and concealed nothing. Last of all, observe, that you do not discover too much, or no more, because all Art hath its Original or Source out of the Spirit of *Mercury*, which is refreshed and raised to Life by the spiritual *Sulphur*, that it becomes Celestial, & with and by the Salt they are made corporal and formal; but the beginning of the Soul, of the Spirit, and of the Body, let it be and remain a Magnet, even as it is, and can be acknowledged to be nothing else. This is the summe in brief, that without the Spirit of

Mercury, which is the only true Key, you can never make Corporal Gold potable, nor the Philosophers Stone. Let it remain by this Conclusion, be silent; for I my self will at present say no more, because Silence is enjoyned thee and me by the orderly Judge, recommending the Execution and further Search thereof to another, who hath not as yet reduced the Matter into a right Order.

Chap. IV.

Of the Spirit of Copper.

The Star of *Venus* is very difficult, and not well to be calculated, as all *Mathematicians* and *Astronomers* will bear me witness; for its course is found to be otherwise than that of the other six Planets, and therefore its Birth is otherwise; for the Birth of *Venus* possesses the First Table, after *Mercury*, as for what concerns the Generation of Metals. *Mercury* makes active, but *Venus* provokes, giving Lust and Desire, together with the Beauty which gave occasion thereunto; though I am accounted no *Astronomer*, nor do I give my self out for one, who knows to calculate the Course of the Heavens; for I should spend my time in my Cell in Prayer, but that the spare hours after my Devotion is ended, may not be spent in vain, I have ordered and proposed it as my aim and intent to exercise my self, and to spend those hours in the knowledge of Natural things. So likewise it is not well to be reckoned what arises, grows or proceeds from *Venus* or whence she arose, grew, or proceeded; for she is superfluously cloathed more than she needs, and yet must want that which she needs most of all in her Constancy.

But you must be advertised, that *Venus* is cloathed with a Celestial *sulphur* which far surpasses the brightness of the Sun; for there is more and more abundant *Sulphur* in her than in Gold; but it requires a knowledge what the Matter of that Gold *Sulphur* may be, which is, and rules so plentifully in Copper, and whereof I make so great a Cry: know then that it is likewise a flying very hot Spirit, which can pass through and penetrate, as also ripen and digest all things, as the imperfect Metals into perfect, which the inexpert will not believe. And here a Question presents it self at hand; *How the Spirit of Copper can make other imperfect Metals perfect, and make them ripe, whereas in its own Body it is imperfect and inconstant?* For Answer, I say as I have often said, that this Spirit cannot possess or inhabit a permanent Body in Copper; for when the habitation is burnt by Fire, the Spirit goes away with it, and must with impatience leave its Lodging, for it dwells therein as a Sojourner; but it hath protection in the permanent fix'd Body of Gold, whence no man can expel it, without the Warrant of an especial Judge; for it is put into the inheritance as an Heir, and taken Root by her permanent Body, that she cannot easily be expelled. The Tincture which *Venus* hath obtained, is in like manner found in *Mars*, more powerful, high and Noble; for *Mars* is the Man, and *Venus* the Woman, which I speak more of, seeing I write of them. This Tincture is delivered in *Verdigreece*, and likewise it is found in *Vitriol*, as in a Mineral whereof a peculiar Book might be wrote. In all these things a combustible *Sulphur* is found, and yet a *Sulphur* which is incombustible, this is a strange thing, one is a white Sulphur, the other is red in the operative generation; but the true Sulphur is incombustible, for it is a pure true Spirit, whereof an incombustible Oil is prepared, and it is the same Sulphur which is made out of one Root from the Gold-Sulphur.

I open many Mysteries, which ought not to be; but what should I do? to conceal all is not answerable, but a measure is good in all things, as you may observe in my last Advice of protestation; forget not my desire therein.

This Sulphur may well be called the *Sulphur of the Wise*; for all Wisdom is found therein, unto the *Mercurial* Spirit; which excels it, which together with the Salt of *Mars* must be put together by a spiritual Conjunction, that three may come into one understanding, and be advanced to equal operations. This spiritual Sulphur proceeds in the same manner and form out of the upper Region, as doth the Spirit of *Mercury*, but in another manner and kind, whereby the Stars manifest a separation in fix'd and unfix'd, in colour'd and uncolour'd things.

The Tincture consists only in the *Spirit of Copper*, and most of all in that of his Bed-fellow; it is a meer Vapour, stinking and ill-sented in its beginning; this Mist must be dissolved in the manner of a Liquor, that the stinking, incombustible Oil may be prepared thereof; but yet it must have and take its beginning out of *Mars*; this Oil unites freely with the Spirit of *Mercury*, assuming all Metallick Bodies speedily unto them, if they be first prepared in all points as I have advised in my Keys.

I observe not the Order of the Planets, and not without just grounds; for I observe the order of their Birth, by which I am directed; for because *Venus* hath much Sulphur, she is sooner digested and ripened together with *Mars*, before other Metals; but because unconstant *Mercury* shewed them both too little assistance, therefore no room is left him to work harder, by reason of the superfluous Sulphur, so that they could obtain no melioration of their unfixt Bodies. Now I will reveal a Secret unto thee, that Gold, Copper, and Iron have one Sulphur, one Tincture, and one Matter of their Colour; this Matter of the Tincture is a Spirit, a Mist and Fume; as aforesaid, which can penetrate and pass through all Bodies, if you can take it, and acuate it by the Spirit which is in the Salt of *Mars*, and then conjoin the Spirit of *Mercury* therewith in a just weight, purging them from all impurity, that they be pleasant and well sented, without all Corrosives, you have then such a Medicine, whereunto none in the world may compare, being fermented with the bright shining Sun, you have made an entrance penetrating to work, and to transmute all Metals.

O Eternal Wisdom from the beginning! how shall we thank thee for such great Mysteries, which the Children of Men do no wayes regard, but are despised by the greater number, to know what thou hast concealed in Nature, which they see before their Eyes, and know it not; they have it in their Hands, and comprehend it not; they deal with it, and know not what they have, nor what they do, because the Internal is concealed. I will yet reveal this unto thee in truth, and by the Love of God, that the root of the Philosophical Sulphur, which is a Celestial Spirit, is found with the root of the spiritual supernatural *Mercury*, as also the beginning of the spiritual Salt, are in one, and found in one Matter, out of which the Stone is made, which was before me, and not in many things, though all Philosophers speak as if the *Mercury*, *Sulphur*, and *Salt* were each one a part by themselves and distinct, that the *Mercury* is found in one, the *Sulphur* in another, and the *Salt* in a third; yet I tell you, this is only to be understood of their superfluity, which is found to abound most in each, and may be used and prepared divers ways particularly with profit, both for Physick and transmutation of Metals; but the Universal, which is the supreamest Treasure of Earthly Wisdom, and of all the three Principles, is one only thing, and is founded and extracted out of one only thing, which can make all Metals into one, it is the true Spirit of *Mercury*, and Soul of *Sulphur*, together with the spiritual Salt, united together, inclosed under one Heaven, and dwelling in one Body, it is the Dragon and the Eagle, the King and the Lion, the Spirit and the Body, which must tinge the Body of Gold to a Medicine, that it may gain power plentifully to tinge his other Companions.

O thou blessed Medicine given by God thy Creator! O thou Celestial Magnet of great attractive Love! O thou valid substance of Metals, how great is thy power, how uninventive is thy virtue, how durable is thy constancy? happy is that man on Earth who knows thy Light in truth, which all the world takes no notice of; he shall not see poverty, no Disease shall touch him, nor no sickness hurt him, till the appointed time of death, and till the last hour predestinated for him by his Heavenly King. It is impossible for all the tongues of Men to utter the Wisdom which is laid in this Treasure of the Fountain, all Orators must be silent and ashamed at it, yea terrified and not able to speak a word, when they shall behold and discern this supernatural Glory, and I my self am afraid when I consider that I have discovered too much. But I hope to prevail with God by Prayer, that he will not charge it on me as a deadly Sin, because I began the Work in his Fear, obtained it by his Grace, and revealed it for his Glory.

O thou holy everlasting Trinity! I praise, honour, and magnifie thee with Heart and Mouth, that thou hast revealed unto me the great wisdom of this earthly World, next unto thy Divine Word, whereby I have known thy Almighty Power, and supernatural Wonders, which Man will not discern; I heartily beseech thee to give me more understanding and wisdom, that I may bestow the use and profit thereof with a continual Sacrifice of Praise before thee, unto the Christian-like Love of my Neighbour, and to my own welfare both spiritual and corporal, in power and virtue, that thy Name may be made glorious, honoured, and praised, for all thy works in Heaven and Earth; and that my Enemies may know, that thou art the Lord full of eternal Wonders, that they may repent and be converted, and not be drowned in the falshood of Darkness. God the Father, Son, and Holy Ghost help me, and all of us, from his heavenly Throne, exalted above all Glory, Might, and Majesty, whose Wisdom hath neither beginning nor end, and before whom all Celestial, Earthly,

Of Natural and Supernatural Things, and Others

and Hellish Creatures must tremble with fear, to him be Glory forever, *Amen*.

O *Seraphin*! O *Cherubin*! how great are thy Wonders and Actions, look graciously upon thy servant, and be entreated to be pacified that he hath manifested this.

The Reader must moreover know concerning the Generation of Copper, and observe, that it is generated of much *Sulphur*, but its *Mercury* and Salt are in an equality, for there is found to be no more or less of the one than of the other, seeing then that the *Sulphur* in quantity excels the *Mercury* and the Salt, thence arises a great coloured redness, which possesses the Metal, that the *Mercury* cannot perform its fixation, that a fixt Body should be generated thereof. Observe and understand it so of Copper, that the form of *Venus* Body is so stated as that of a Tree, which abounds in Rosen, as the Larch Tree, the Firr, the Pine, Deal Tree, and other sorts of Trees more, the Rosen of the Tree is its *Sulphur*, which it evacuates at sometimes by reason of its superfluity, for it cannot bear it all; such a Tree which is tinged with abundance of fatness, by the digestion of Nature and the Elements, burns quickly and freely, and is not ponderous, nor so durable as is the Oak, or other hard wood which is close and compact, whose Pores are not so open, as those sorts of light wood, and wherein the Sulphur doth not so predominate, but the Oak hath therefore the more *Mercury*, and a better Salt than the Pine, Firr, and Deal trees have, and such wood doth not float so well above the water, as the Deal, being bound & closed up compactly, so that the Air is easily prevented in bearing it up. So is it to be observed of Metals, and especially of Gold, which by reason of its abundant, fixt, digested and ripe *Mercury*, hath a very close, fast and compact, fixt and invincible Body, which neither Fire nor Water, Air, nor any Corruption of the Earth can prejudice, that the consuming power of the Elements can do them no harm; this fixedness & close compacted Conjunction gives evidence of its natural ponderosity, which cannot be evidenced in other Metals, which is to be observed, not only by weighing it in the scales, but likewise you will find it thus: if you lay but a scruple of pure Gold upon a hundred weight of Quicksilver, it immediately sinks to the bottom, whereas all other Metals being laid upon Quicksilver in like manner, float on the top of it, and sink not to the bottom, because they are more open, that the Air or Wind can penetrate them and bear them up.

Now what further concerns the Spirit of *Venus* or Copper in Physick, you must last of all take notice and observe, that it is throughout in its virtue and power discerned to be very wholsom and beneficial, not only that Spirit which lies in the first *Ens*, but also that very Spirit which is found in the last Matter, its virtue, power and operation is, that it is preferred before all other Medicines in the Rising of the *Matrix*: It's like is not yet found particularly against the *Falling Sickness*. This Spirit hath also received an especial gift to dry the *Dropsie* up; it preserves the Bloud from putrefaction, digests all which is adverse to the Stomach, breaks the Stone, of what kind soever it be. Externally in Wounds, this Spirit lays a ground to heal: *Noli me tangere* and all other Sores cannot defend themselves, nor their ill Qualities, but this Spirit doth assault them, and prepares a good ground for their Cure; externally it mundifies and searches out that whereby the Medicine may operate, fasten, and make a beginning of the Cure. Internally this Spirit penetrates through & through, searching out all that is evil in the Body; even as doth the noblest Vulnerary potion; No Imposthume can withstand this Spirit, but is reformed by it. I say briefly, observe the Spirit of *Venus* very well, it will manifest it self to admiration both internally and externally, that many will esteem it to be incredible & supernatural. Last of all, you must understand that this Spirit of *Copper* is a fiery Spirit, penetrating, searching and consuming all evil Humours, and superfluous Flegme in Man and Metals, and may in reason be accounted the Crown of Medicines; it is very fiery and sharp, incombustible, but spiritual and unformal, and therefore as a Spirit it can particularly help to make unformal things fiery, digest and ripen them; and if you are a true Naturalist, I recommend this Spirit unto thee; it will not fail thee in the least, in any necessity of Health or Wealth, in case you observe it rightly, and execute according to Justice. I hope my Call and Request will at last take place, and have a hearing with those who regard Nature, and have an earnest and longing desire to search out, and learn, whereby they may whet their Wits, open their Eyes, and let their Ears hear, and learn such a thing out of my Advice, which was never taken notice of, or learn'd before, and is to be found in this Spirit of Copper, internal and external. He that doth not observe, or truly understand my Writings, will not fathom many Secrets, nor search out to purpose and in truth, nor learn to advantage without me, therefore no Man can direct me, as concerning the Spirit of Copper, except he hath beforehand inverted and turned the Copper inside outwards, and truly learned all the Mysteries of its internal Virtues, as I have done, if he can find out any thing better, which I know not, I earnestly desire him not to conceal any thing, so shall his instruction be well rewarded, with a thousand-fold advantage, and recommend you herewith to the Highest Creator.

> *Vain Reason cannot alwayes apprehend*
> *Each matter which* Venus *can bring to an end:*
> *No man can find it presently in sence,*
> *Vain Reason banns it far away from thence;*
> *Such a Spirit only can all things speed,*
> *So that* Mercury *be joyn'd with it indeed.*

Chap. V.

Of the Spirit and Tincture of Mars.

Mars and *Venus* have a Spirit and Tincture as well as Gold and other Metals, be that Spirit which is in each Metal never so mean and little. It is undeniable and known to all, that many men have many minds, though all men originally are of one first Matter, born and produced from one Seed; yet have they divers different Minds, because the Stars have so operated them, and not without cause; for the influences of the great World operates the next to it in the little World; for all Opinions, Natures and Thoughts, together with the whole complection of Man proceed alone from one Influence of the Stars, manifesting themselves according to the Course of the Planets and Stars, so that nothing can prevent, nor can such Influences hinder it, when the Birth hath attained to the end of its perfection.

As a man is naturally inclined to study; one delights in Divinity, another in the study of the Laws, a third in Physick, a fourth will be a Philosopher; moreover there are many Wits who are naturally inclined to the Mechanicks; as the one is a Painter, another a Goldsmith; the one a Shoomaker, the other a Taylor, a Carver, and so forth, divers and innumerable; all this happens by the Stars influence, whereby the Imagination is supernaturally founded & fortified, and whereupon it is resolved to rest; as it is found, that what a man hath once conceived in his Mind, and framed a foundation thereof, none can divert him from a constant resolution and relying thereon, except Death, which at last concludes all. So is it to be understood of *Alchymists,* who are set upon the search of Natures Secrets, they intend not to cease, till they have discover'd Nature, absolved it quite, and brought all to an end, which cannot well be done.

Even so is it to be understood of Metals, according as the Influence and Imagination is from above, so is the Form; and although the Metals be called Metals in general, and are such, yet you have understood by the various minds of men, which yet proceed from one Matter, that there may be manifold and divers Metals, one hot and dry, another cold and moist, a third assuming a mixt Nature and Complection to it self. Therefore the Metal of *Mars* being ordained in its degree by a gross Salt before others in the greatest quantity, is found to have the hardest, ungentle, strongest, and grossest Body, which Nature appropriated and granted to it, it hath the least portion of *Mercury*, but more of *Sulphur*, and most of *Salt*, hence, and from such a mixture or composition is its corporal essence descended, and born into the world by help of the Elements. Its Spirit is like to the other Spirits in operation, but if you can know the right and true Spirit of *Mars*, I tell you truly, and in true Wisdom, that one grain of its Spirit or Quintessence drunk with the Spirit of Wine, strengthens the Heart, Courage, and Senses, so that you shall fear no Foes; it raises up in him the Courage of a Lion, and provokes a desire to hunt and fight at *Venus* sports. When the Conjunction of *Mars* and *Venus* are rightly placed in a certain Constellation, they bring Fortune and Victory in Love and Affection, in Battel and joy, remaining in unity though the whole World should be against them: But because I am an Ecclesiastick under Church Government, and dedicated my Soul to God, without provocation of humane desires, and lusts of the Flesh, for they lead a direct way to Hell without leave; but Gods Commands, Fear, and a rejection of Mans Will, which are tollerated by his Commands, prepare a way to Heaven, if they continue in the true calling upon, and in the true and right Faith of the only Throne of Grace, Mediator and Patron *Jesus Christ* our Saviour. All Martial Diseases are expell'd, cured, and healed in an admirable manner by this Spirit; such as are the *Bloody Flux*, the Disease or Menstruous *Fluxes* of Women, both white and red, and all other *Fluxes* of the Belly, and open *Sores* in the Legs, or any part of the Body, together with all those Diseases, both internal and external, howsoever they are called, which bloody *Mars* hath caused, which I omit to nominate particularly, being well known unto the discreet Physician what Diseases are subject to the jurisdiction of *Mars*. If the Spirit of Iron be truly known, it hath a secret affinity with the Spirit of *Venus*, so that both may be conjoined in one, both becoming one only matter, of a like operation, form, substance and being, healing and expelling the self-same Diseases, as also to bring the particulars of the Metals into a change with profit, praise, and excess. But properly *Mars* must be observed thus with its virtues, that in his Corporal form he only hath an earthly Body, which may be used in many things, for to stanch Bloud, externally in Wounds, to graduate *Luna*, internally to stop or bind the Body, which yet is not good at all times, and may be used both internally & externally in mans Body, as likewise in Metallick affairs; because without the true known means, which Nature hath in her secret Closet, much profit cannot be gotten *per se*.

One thing more I must at present propose, that the Magnet and true Iron perform almost a like benefit in Corporal Distempers, having almost one kind of Nature in and with them, as it is with it in the Celestial, spiritual, and Elementary Intellect, between the Body, Soul, and the Chaos, out of which the Soul and Spirit went, the Body at last was found out of the Composition.

How shall we now do? the gross dull-witted Lads will not apprehend it, the middle sort of Wits will take no notice of what I write, and the supernatural wits will descant too much upon it; I must find out a remedy, and would willingly preserve all these over-wise-people to be my Friends still. I will now teach, instruct, and presently inform you, seeing that the Argument it self declares and pronounces its definitive sentence, therefore the resolution lies open, and can be declared and resolved, reserved nor directed to any other sentence of the understanding, further than for it self.

Last of all, reserve this hereupon in this Chapter, that there can be no House kept to stand in unity between the Married Couple, if the one of them turn his Coach and drive to the East, and the other towards the West, for they are not equal, so that they cannot draw the Coach together in an equal weight, whereby there arises a great dissention and hinderance, in obtaining that which was intended: but if true Married People will carry on their House-keeping with a right subsistance, they must be of one spirit, mind, judgment and virtue, to accomplish all whatsoever is in their heart and mind, and that the one operate into the other, if their Love and Truth shall be permanent; for want of one of these things, the three principles cannot be truly together; for the *Mercury* is banisht, and too little by reason of the firmness and constancy; the *Sulphur* is too little, it cannot warm the Body of Love, because it is very much extinct; the *Salt* likewise hath not its right, convenient, natural kind, but is too hard and too much, seeing it makes a hard coagulation, is sharp and biting, because it doth not manifest it self in truth and constancy. Even so it goes now in the World, which goes astray, and is pregnant with such Vices, for the constancy is but small, the Love little, and Truth as little.

I hope you will take this Philosophical Example in good part, because *Syrach* doth both praise and dispraise the goodness, truth, and wickedness of a false Woman, and both after a different manner; and herewith I bid *Mars* Farewell, saying, that no man knows how to distinguish the Sentence of one, much less of all things, but he who hath in this point taken notice of them, learned and experimented their Nature and Properties, and truly known and discovered them. God our Heavenly Father, the Everlasting Power, proceeding from all beginning, separate us so in the Form, that the terrestrial corruptible Body may again attain unto, contain, and receive the Celestial, Spiritual and Incorruptible Revelation. *Amen.*

> *Maist thou not know me alone indeed,*
> *And procure a pure help for me in need;*
> *Resolve then, and hear what I do speak or say,*
> *So shalt thou find what I can do for aye.*

Chap. VI.

Of the Spirit of Gold.

The Clearness of Heaven hath now commanded me to govern my Pen, to reveal a matter of valour and of permanency; for the Sun is a burning and consuming Fire, hot and dry, wherein is concealed the right and true virtue of all Natural things; this virtue of the Sun worketh Understanding, Riches, and Health. My Mind is very much grieved, and my Spirit is terrified within it self to discover it publickly, which was not publish'd in common before, and to make it vocal, which was concealed in the deep with great secresie. But if I consider in my self, and enter into my Conscience, I could find no alteration, nor catch at any thing to disturb my mind, or bring it to another resolution, which might cause many Obstructions: Yet will I speak with discretion, and write understanding, that no evil with may follow, but rather that I may gain a grateful profit, which I have pourtrayed after the manner and occasion, as the Philosophers before me have done.

Mark now, give your Mind perfect Thoughts, refrain all strange matters, which are not serviceable to your speculation of Philosophy, but rather cause a ruine of that benefit, which you pursued with so much diligence; and know if you have a hearty desire and strong affection to gain the Golden Magnet, that in the first place your prayer be truly directed to God, in true Knowledge, Sorrow, Repentance, and true Humility, to know and learn the three distinct Worlds which are subject to Humane Reason; as, there is the Super-celestial World, wherein the right immortal Soul hath its seat and residence, together with its first coming, and is according to Gods Creation the first moveable Sense, or the first moving sensible Soul, which hath operated the Natural Life from a Supernatural Essence; this Soul and Spirit is at first the Root and Fountain, the first Creature which arose to a Life, and the first Mover, whereof there hath been so much Disputing among the Learned.

Now take notice of the second Celestial World, and observe it diligently, for therein the Planets rule, and all the Stars of Heaven have their course, virtue and power in this Heaven, performing that Service therein whereunto they are by God ordained, and in this service they operate the Minerals and Metals by their Spirit.

Go now out of these 2 distinct Worlds into the third, wherein is contained and found what the other two have wrought, to wit, the Super-celestial and the Celestial worlds; out of the Super-celestial arises the Fountain of Life, and of the Soul; out of the other Celestial world the light of the Spirit; and out of the third or Elementary world, the invincible Celestial Fire, which yet may be felt, out of which, that which is tangible is digested; these three Matters and Substances produce and generate the Form of Metals, among all which Gold hath the pre-eminence, because the Sidereal & Elementary Operation hath digested and ripened the *Mercury* in this Metal the more perfectly to a sufficient ripeness.

And even as the Male-Seed is injected into the Womb, and touches the *Menstruum*, which is its Earth, but the Seed which goes out of the Male into the Female, is operated in both by the Sydereal and Elementary, that they be united, and nourished by the Earth unto the Birth.

Even so understand it likewise, that the Soul of Metals which is formed and conceived out of the *Chaos* by an intangible, invisible, incomprehensible, concealed, and supernatural, Celestial Composition of Water and Air; afterwards it is further concocted by the Celestial Elementary Light and Fire of the Sun; whereby the Stars move the Powers, when its heat is perceived in the inward parts of the Earth, as in the Womb, for the Earth is opened by the warm operating property of the upper Stars, that their infused Spirit yield a nourishment unto the Earth, that it may bring forth somewhat, as Metals, Herbs, Trees and Animals; where each one in particular brings its Seed with it for its farther augmentation and encrease: And as hath been mentioned, even as Man is begotten spiritually and heavenly, Soul and Spirit, and by the nourishment of the Earth in the Body of the Mother is formally brought up to perfection; even so, and in like manner, is to be observed and understood of the Metals and Minerals in all points.

Of Natural and Supernatural Things, and Others

But this is the true mystery of Gold, which I will make good to you by an Example and Parable to certifie you, whereby the possibility of Nature, and its Mystery is to be found after this manner.

It is evident, that the Celestial Light of the Sun is of a fiery Quality and Essence, given unto it by a Celestial, fixt and permanent sulphurous Spirit, by the most High God, Creator of Heaven and Earth, to preserve its substance, form, and body; which Creature, by its swift motion and course, is enflamed and kindled by the Air through that swiftness in a continued manifestation of it; this inflammation can never be extinct, nor decay in any of its power, so long as its Course last, or this whole Created visible World shall remain and continue, because there is no combustible matter at hand which is given unto it, by whose consumption this great Light of Heaven should fall to decay.

Even so Gold is so digested, ripened, and made into such a fixt invincible Nature by the Superiours in its Essence, that nothing can hurt it in the least, because the superiour Stars have past through the inferiour, that the inferiour fix'd Stars by the influence and donation of the superiour, cannot in the least give place to its like, for the inferiour have obtained such a fixedness and permanency from the superiour; this you may well retain, observe, and take notice of as concerning the first Matter of Gold.

I must yet produce one comparison according to the Philosophical custom, of the great Light of Heaven, and of that little terrestrial fire here daily kindled, and made to burn before our Eyes; because that great Light hath a magnetick simulation and an attractive living power with the small fire here on earth, but yet it is unformal and incomprehensible, only it is found to be spiritual, invisible, insensible and intangible.

It is to be observed and remembred, as experience manifests, and is proved, that the great Light of heaven bears an especial sympathy, affection and inclination to the little earthly fire, by means of the spiritual Air, whereby they are both promoted and preserved from Mortality; for behold, when the Air receives into it a Coruption, by too great humidity attracted up by it, that Clouds are generated by Mists, and farther coagulations, which hinder the Sun-beams that they cannot have a reflection, nor get a right penetrating power. So likewise the small terrestrial fire doth not burn so lively in dusky, dark, rainy weather, nor manifests it self with joy in its operation, as it doth when there is a fair, pure, serene, unfalsified heavenly Air; the reason is, because the sympathy is bound and hindered by the obstruction of those Accidents and the waterish Air, so that the attractive power is grieved, that it cannot accomplish its compleat Love and Operation as it should, for this hinderance brings the aquosity to the contrary Element.

Now even as the Sun, the great Light of Heaven, hath a peculiar community and sympathy with the small terrestrial fire to attract unto it, after a Magnetical manner; So also the Sun and Gold have a peculiar understanding, and an attractive power and sympathy together; for the Sun hath wrought the Gold by the three Principles, which have their Magnets, being nearest related to the Sun, and hath gained the next degree to it, for that the three Principles are found to be most mighty and powerful therein, Gold immediately succeeds it in its corporal Form, being composed of the three principles, and hath its beginning and off-spring from the Celestial and Golden Magnet.

This is the supremest Wisdom of this world, a wisdom above all wisdom, yea a wisdom above all Natural Reason and Understanding; for by this wisdom is comprehended first of all Gods Creation, the heavenly Essence, the Firmamentary Workings, the spiritual Imagination, and the corporal Essence, it contains all qualities, and properties, and all whatsoever sustaines and preserves Mankind. In this Golden Magnet sticks and lies buried the resolution and opening of all Metals and Minerals, their domination, as also the first Matter of their generation, their power over health; and again, the coagulation and fixation of Metals, together with the operation of expelling all Diseases: Take notice of this Key, for it is Celestial, Sydereal and Elementary, out of which the terrestrial is generated, it is both Supernatural and Natural, and is generated Celestially of the Spirit of *Mercury*, Spiritually of the Spirit of *Sulphur*, and Corporally of the Spirit of *Salt*; this is all the way, the whole Essence, the beginning and end; for the Spirit and the Body are bound up together in one by the Soul, that they can never be separated, but produce a very perfect, durable Body, which nothing can hurt. Out of this spiritual Essence, and out of this spiritual Matter, out of which first of all Gold was made into a Body, and became corporal, out of it is made a more true and compleat *Aurum potabile* than out of Gold it self, which must first of all be made spiritual, before a potable Gold can be prepared out of it.

This Spirit cures and heales the *Leprosie* and the *French Pox*, as being an over-fix'd Mercurial Essence, dries up and consumes the *Dropsie*, and all running and open Sores, which have raged a long season, it strengthens the Heart and Brain, makes a good Memory, generates good Blood, brings Lust, Delight and Desires in humane incitation unto Natural Affections. If the Quintessence of Pearl be mixt with the Tincture of Coral, and be administred with an addition of an equal quantity of this Spiritual Essence of Gold, the Dose of two grains taken at once in a just observation, you may be bold & confident of the truth, that no disaster of any Natural Distemper can harm you, or happen to you, to the prejudice of your health, because the nature resides only in the Spirit of Gold, to alter, remove and amend all weaknesses, so that the Body shall be adjudged perfect and free from any Disease. The Quintessence of Pearl corroborates the Heart, and make a perfect Memory, of the five senses. The Tincture of Coral expels all poison, and evil Spirits which fly from the good. So can the Soul of Gold in a Water turn the spiritual Essence of the Pearl, and the Sulphur of the Coral united in one, perform such a thing which otherwise Nature could not be intrusted with, but seeing that Experience hath manifested it, and confirmed the undeniable Truth, therefore this Cordial in this temporal Life is, and ought in reason to precede all other Cordials with admiration and admirable Effects, be they called by what name soever. I am an Ecclesiastical person, obedient to the Ecclesiastical degree, related to the *Benedictine* Order by a Spiritual and Divine Oath, by which Order with my internal Prayer, I obtain comfort and promises of Gods Word, a refreshment to my Soul, but in a corporal temptation of my weaknesses, and for my Brethren I have not found and used a better corroboration by Gods Blessing, than these three Compounds united: God give, bless, and increase this Virtue and Power unto the End of this temporal World, which Man must change together with Death. O thou golden power of thy Soul! O thou golden intellect of thy Spirit! O thou golden operation of thy Body! God the Creator keep thee, and grant

Of Natural and Supernatural Things, and Others

unto all earthly Creatures, who love and honour him, the true understanding of all Gifts, that thy Will may be done in Heaven and on Earth: This is enough revealed at present concerning the Spirit of Gold, until the coming again of *Elias*.

Hereunto I add a short process:

Take a Spirit of Salt, therewith extract the Sulphur of Gold, separate the Oil of Salt from it, rectifie the Sulphur of Gold with Spirit of Wine, that it be pleasant without Corrosive; then take the true Oil of *Vitriol*, made of the *Vitriol* of *Verdigreece*, therein dissolve *Mars*, thereof make a *Vitriol* again, and again dissolve it into an Oil or Spirit, which rectifie in like manner as before with Spirit of Wine, conjoin them, and abstract the Spirit of Wine from it, resolve the Matter which remaines dry in Spirit of *Mercury*, according to a just weight, circulate and coagulate it when it is fix'd and permanent without Ascention, you have then a Medicine to tinge Man and Metals, if it be fermented with prepared Gold.

Chap. VII.

Of the Spirit of Silver.

The Tincture and spirit of Silver manifests its Colour of a Watchet or Sky-colour, otherwise it is a waterish Spirit, cold and moist, not so hot in its degree as that which is found in *Gold*, *Mars*, and *Venus*; for *Luna* is more phlegmatick than fiery, though it be brought by the Fire out of its waterish Substance into a coagulation; and even as the Metals gain their tinging Spirits and Coagulation, in like manner do stones get their fixedness, and colour, as out of one Influence. A fix'd coagulated *Mercury* is found in the *Diamond*, therefore it is fixeder and harder than the other stones, and cannot be so broken; so the tincture of *Mars*, or the Sulphur of Iron is found in the *Ruby*, the Sulphur of *Venus* in the *Emerald*, the Soul of *Saturn* in the *Granate*; in Tin the tincture which is found in the *Topaz*; and *Crystal* is appropriated to common *Mercury*; in the *Saphire* is found the Sulphur and Tincture of *Luna*, but each one according to a peculiar understanding, and according to its kind, and in Metals according to their form and gender; for when the blew Colour is taken and extracted out of the *Saphire*, its Rayment is gone, and its other Body is white as a *Diamond*, wanting only the hardness that is in a *Diamond*; even so when Gold hath lost its Soul, it yields a fix'd white Gold Body, which by searching Students and young Artists is called fix'd *Luna*.

Wherefore you must now understand and observe, that even as I have declared unto you concerning the *Saphire*, for your apprehension, even so on the other side, you must learn to what purpose my Speech is intended, for your Instruction concerning Metals.

For this blew Spirit is the Sulphur and the Soul, whence the Silver receiveth its Life, both in and above the Earth, by Art, and the white Tincture of the Silver upon white stands in the Magnetick form of an everlasting thing, or Creature, wherein is likewise found the first *Ens* of Gold.

Of Natural and Supernatural Things, and Others

O ye high qualified Orators! where is your voice in this case to explain this Mystery? And you conceited Naturalists, where is your Writings and Advice of Reason? And you Physicians, Whither is your Opinion flown, to fetch somewhat afar off over the Seas for to cure the *Dropsie*, and all *Lunary* Distempers? You will say, that this my speech is too dark for you; is it so? then kindle the terrestrial Light, seek, and be not ashamed to make acquaintance with *Vulcan*; and let nothing be irksome unto you, so will you find by the assistance of the Eternal God, that the Spirit of *Silver* contains in it to cure and expel the *Dropsie* quite alone, as the Spirit of *Gold*, and as that of *Mercury* can expel the *Consumption* radically, or in the root, even so that the Center of those Diseases cannot be found any more. But that *Luna* in the veins of the Earth is not furnished with such a hot substance or quality in its degree, but is subjected to a Waterish Nature; this fault lies upon that great Light of Heaven, which by reason of its Waterish influence, hath implanted such qualities in the other Creatures, and Planets of the Earth, than it hath in *Silver*. And albeit that *Silver* contains a fix'd *Mercury*, which is generated in it, yet it wants a hot, fix'd *Sulphur*, truly to dry up and consume its Phlegme, whereby it hath not obtained a compact Body, unless it be done afterwards by the art of the Little World. And seeing that its Body is not compact by reason of the abounding watery substance, its Pores therefore are not rightly defended, nor closed to undergo the weight and endure a Battel with the Enemies; all which Virtues are to be found in *Gold*, if it shall overcome all Enemies, and endure all trials without defect.

All things are difficult in the beginning, but when they are brought to an end they are easie to be understood and apprehended. If you do truly observe the Spirit and the Soul of *Luna*, and learn to know it truly, you may quickly compass the midst of the Work, how it shall afford the end with profit; wherefore I will now propose to you an Example, and instruct you by a Countrey-Rule, that you may apprehend it, and consider of it, as Childrens Play, in a high and weighty Matter, that you may search it out with advantage; as followeth:

A common Peasant casts forth (or sows) his Seed in a Field well dunged and prepared, this Seed after putrefaction, sprouts forth of the Earth by the operation and furtherance of the Elements, and sets before our Eyes the Matter of Flax together with its Seed which it brings with it augmented; this Flax is pluck'd up, and separated from its Seed; but this Flax cannot be used and prepared for any work profitably, except it be first putrefied and rotted in water, whereby the Body is opened, and gains an ingress of its doing good; after this putrefaction and opening, it is again dried in the Air and Sun, and by this coagulation it is again brought into a Formal Being, that it may do future service. This prepared Flax is afterwards buck'd, beaten, broken, peel'd, and last of all dress'd, that the pure may be separated from the impure, the clean from the filth, and the fine from the course; which otherwise could not be done at all, or brought to pass without the preceding preparation; this done, they spin Yarn of it, which they boil in water over the Fire, or else with Ashes set in a warm place, whereby it is purified afresh, whereby the filth and superfluities are fully separated from it, and after a due washing the Yarn is dried again, delivered to the Workmen, and Cloth weaved of it; this Cloth is purified or whitened by a frequent casting of water upon it, cut in pieces by Taylors, and other people, so converted to future services in houshold affairs, and when this Linnen is quite worn out, and torn, the old Rags are gathered together, and sent to the Paper-Mills, whereof they make Paper, which is put unto divers uses.

If you lay Paper upon a Metal or Glass, kindle and burn it, the vegetable *Mercury* comes forth and flies away into the Air, the Salt remaines in the ashes and the combustible *Sulphur* which is not so quickly consumed in the burning, dissolves to an Oil, which is a good Medicine for dim and defective Eyes. This Oil hath in it a great fatness, which is the Matter of the Paper, contained originally in the Seed of the Flax; so that the last Matter of the Flax which is Paper, must again be dissolved into the first Matter, which is the fat Sulphurous Oyliness of the Flax-seed, together with the separation of its *Mercury* and *Salt*, that so the first may be made of the last, and the ground-work revealed, so the Virtues and Operations known by the first.

And though this Discourse be gross and not subtil, yet you may learn thereby to know what is subtile and secret; for that which is subtile must be infused into the ignorant by course Examples, that thereby they may be taught to reject the gross, and to embrace that which is subtile.

In like manner understand, that the first Matter of Metals must be observed, known, and found out by the revelation of their last Matter, which last Matter, as there are the perfect Metals, must be separated and divided asunder, that it may plainly appear singly before the Eyes of men. Out of which separation may be judged and learnt what the first Matter was at the beginning, out of which the last was made. Accept of this Advice concerning *Luna* at present. I could have said more, but I must desist at this time until another opportunity; and intreat you heartily, admonishing you by your Conscience, that you observe all that which I have revealed unto you, of all those Letters which are contained in the middle between *Alpha* & *Omega*, & that you keep all the Speeches & Writings, that you may not undergo a denial of pardon for your Sins, & a continued perpetual Vengeance for Eternity; which I at last reveal unto you thus:

Take the Sky-coloured Sulphur extracted out of *Silver*, rectified with Spirit of Wine, dissolve it according to its Quantity in the White Spirit of *Vitriol*, and in the sweet-sented Spirit of *Mercury*, coagulate them together by the fixation of the Fire, you have the White Tincture in your Hands with all its Medicines; but if you can get all their *primum Mobile's*, it is then needless, because you can perfect the Work at once.

Of Natural and Supernatural Things, and Others

CHAP. VIII.

Of the Soul or Tincture of Tin.

Good *Jupiter* possesses almost the mean or middle place between Metals, it being not too hot, nor too cold, not too warm, nor too moist, it hath no excess of *Mercury*, nor of Salt, and it hath the least of Sulphur in it; it is found to be white in Colour, yet one exceeds the other in the three Principles, as it is evidently found in its dissection, the right and true discovery of Nature. It is generated of such a composition and mixture of the three first Principles, being operated, coagulated into a Metal, and brought to the ripeness of perfection. *Jupiter* is a God of Peace, a Lord of Goodness, a Ruler and Possessor of the middle Region; as concerning its State, Essence, Function, Virtue, Form and Substance; for it holds the mean; no special Disease can happen, that *Jupiter* should cause any remarkable damage, if its Medicine be used a little at once, not too much in quantity; it is likewise thought needless, where its Medicines are not required, that they should be administred in strange cases with a just Call, but we should rather abide by those, where the Body and its Disease have an equal temper with the superiour Stars and their assistance, in vertue, power, and operation, and so accord together in their juncture, that there is not found the least contrariety in the Operation, nor in the Operative Nature.

Jupiters Spirit is found not to be wanting in the least, in the generation of Metals, as likewise no one Spirit of all the Metals can be set backwards, because of necessity they accord together from the lowest to the highest degree, and must agree together, as a Metal is perfect in the great Earth, so should the transmutation & augmentation succeed in the little world; understand it after this manner, that all the degrees from the meanest to the highest Metal must be passed through in all perfection, even as the Metals must finish their course, from *Saturn* unto *Gold*, as concerning the permanency of Colour and Body, notwithstanding that *Saturn* possesses the highest place in the highest Region, wherein the stars reign and perform their Course.

Of Natural and Supernatural Things, and Others

The generation of Tin in and above the Earth, is brought to light even as Man is and other Animals, which are originally nourished and fed by the Mothers Milk; there is no Diet to be found on Earth more fit for the nourishment of all men than Milk; for its best part is chiefly an Animal *Sulphur*, which yields the Nourishment. Even in like manner *Tin* is nourished by its Metallick *Sulphur*, which likewise feeds it with the greatest acceptation, it assumes in and to it more heat than *Saturn*, therefore is *Jupiter* more digested & broiled, whereby its Body likewise is more fixt and permanent in the degree of Salt.

He causes in his Dominion and Reign, that good Rule be observed, and Justice done to all men in his Court. The Spirit of Tin is a Preserver from all Distempers & Accidents whereby the Liver is consumed or put into malady; its Spirit is naturally to be compared unto Honey in Taste, its *Mercury* being made volatile, gains a venomous quality; for it purges violently, and penetrates through by force, therefore it is not always to be advised, that its opened *Mercury* should be used alone and simply, but if a Correction precede, there may an excellent benefit succeed, being used in those distempers and diseases, which are immediately subject to its Influence, that is, when its venomous volatility is taken away, and set in a better and fixeder state, which resists the poison.

The Vulgar Physician cannot understand this Description; for this Art and knowledge proceeds not from the bare Talking, but from Experience; the common Physician hath the foundation and egress in speaking, but our Preparation hath its Rise from speaking, and then its foundation first of all out of a certain trial, which manifests it by Experience, and this is firmed upon hard Rocks by manual Operations, but the other stands upon moving Reeds & Sand; wherefore in reason that which is strong and immoveable, made by Natures hand, ought to be prefer'd before bare Speeches, which proceed only from an inconstant phantastical speculation, because the Work always will praise the Master.

At present I do not indeed speak according to my own Poetical manner, nor after such a way as I directed my stile, when I treated of the wonderful generation of the seven Planets in my occult Philosophy, nor after a Magical or Cabalistical manner and custom;

much less do I observe the method which teaches, and diligently marks the Mystical, Secret and Supernatural Arts, to wit, of *Hydromancy, Æromancy, Geomancy, Pyromancy, Nigromancy*, and the like: But my present purpose and intent is directed to reveal Natures Secrets, that all the Lovers of Art, and the Children which seek and desire wisdom, may by Gods Grace, Blessing, and Permission, easily understand, observe, mark, and likewise after diligent observation learn, & retain something that is beneficial; this concerns the generation of Metals in two parts, in the great and in the little World, as likewise what is the true Medicine contained in the inward part of those Metallick and Mineral Forms, which must be apprehended and made moveable by their dissection that their first beginning may be made notoriously visible in three distinct things; Then is Nature stript, and her secret parts discovered by laying off her temporal Cloathing, and all the secret Virtues, Powers, and, Operations revealed for Mans Health. My Persecutors, and, indiscreet Physicians will now tell me, thou talkest much of Geese, and knowest not a Duck; who knows whether all what thou writest be true? I will stick where I am, and remain by what I have tried, and bears the sway among all my Associates and Physitians; so shall I not be deceived, and am assured that I shall not need to take paines to learn any new Matter. He that is of such a resolution, may remain with the Ducks; for he is not worthy of a roasted Goose, nor to learn what is concealed in Nature.

But this in truth I acknowledge, and confess it before the Supreme Trinity, speaking it to the hazzard of that most Noble Ecclesiastical Jewel, that all what I have wrote, and yet shall write in this point, is all true, and shall be found to be no otherwise in truth: But that every ignorant, or vulgar person, which are haters and persecutors of this Mystery, do not well, fully, and clearly understand my Writings at first; alas! that cannot I help; pray unto God for his Grace, and ye Persecutors for pardon, labour without repining, read with understanding, then will no Mystery be withheld from you, but will be very easie for you to find out. I moreover admonish, that the finder of this gift of God, above all things give thanks unto God day and night without ceasing, with all reverence and due obedience, from the bottom of his heart; because no Creature can yield sufficient praise which may recompense so great a benefit; but Diligence is known by a right and true industry according to our capacity. I have done my part, which I hope to justifie before God and the World; for what my Eyes have seen, my Hands felt, and apprehended by an undeceived Judgment, that shall no man take from me in this Life; only Death, which is the determiner of all things.

This my Speech hath indeed had no force to poure forth from it what is written by me herein; but what I have done is not out of curiosity, nor out of a desire of vain and transitory Glory; but I have been induced thereunto by the Command of Christ the Lord, that his Glory and Goodness in eternal and temporal Matters, should not be concealed from any man, but to the praise, honour, and glory of his holy everlasting Name, that it might be exalted, acknowledged, and revealed in his Majesty by reason of his Highness and Almightiness, through the confirmation of his wonderful Deeds! And secondly, I have been led thereunto by Love and Charity towards my Neighbour, for his good as for my own, and to heap burning Coals on my Enemies heads. And last of all, that all Opposers may know, what erroneous waies others have gone against me, and whether I am most of all to be condemned, or they adjudged most just in what hath been written most truly of the concealments of Nature; & likewise that the supremest Mystery may not quite be suffocated in darkness, nor be drowned in overflowing waters, but be delivered out of the deep and filthy mire of the Ideotish Crew by the right appearance of the true Light, and obtain many witnesses by the spreading abroad of a sure, true, and right Confession, who may follow me in the Writings of Truth. In my Nativity of the twelve Signes in the Zodiack, *Sagitary* and *Pisces* were allotted unto me; I was born under *Pisces*, for I was in Waterishness before my Life, but *Sagitary* set an Arrow to my Heart, whereby I lost my Waterishness, and by the heat I became worthy of the dry Earth; and although at the first the Earth was turned by the Water into a soft substance, yet you must understand that the Water was consumed by the heat of the drying Air, so that all the soft Matter of the Earth went away, and by this drying up was dignified with a Hardness; whereby thou Learner, and much Understander should carefully observe and take good notice, that Tin is subject to all the four Elements, as also to the other principal Planets; which Elements received their Center from above, and are generated as others.

To conclude, I let you know, and give you to understand, that if thou extract out of Benevolent *Jupiter* its Salt and Sulphur, and lettest *Saturn* flux well with it, *Saturn* assumes a fixt body unto it, purges it self, and becomes clear thereby, there being a full change and real transmutation of Lead into good Tin, which may be found to the height by a durable infallible proof. And though you may think this to be false, yet you must take notice, that seeing the Salt of *Jupiter* only by its Sulphur is made more corporal, yet likewise it hath obtained an efficacy and power to penetrate *Saturn*, the basest and most volatile Metal, and bring it to a melioration of its Equals, as you will find it in reality.

Chap. IX.

Of the Spirit of Saturn, *or Tincture of* Lead.

Saturn to generate his Metal Lead, is placed in the upper Heaven above all Stars, but he possesses the lowest and vilest degree in the under-parts of the Earth, even as the supreme Light of *Saturn* is mounted aloft in the highest supremacy of all the Celestial planets, so hath its Children of the lower Region succeeded it in Kind; and Nature hath permitted that *Vulcan* should conduct them to their like, if *Saturn* be content; for the upper light gives occasion thereunto, having generated an unfixt Body of *Saturn*, penetrated with open pores, that the Air can pass through this *Saturnine* Body, that the Air can keep it aloft, but the fire can quickly assault it, because the body is not compact by reason of its unfixedness, so that it must decay, which must be in all points observed by him that will attain to the search of it; for there is a great difference between the fix'd and unfix'd bodies, and of the causes of their Constancy and Inconstancy. And though *Saturn* hath an especial ponderosity above other Metals, yet observe, when they are poured forth together, after their union in the Flux, the other Metals alwaies settle at the bottom, even as it likewise comes to pass in the pouring of *Antimony* through with other Metals, whereby it is evident, that the other Metals fall through equally, and are more compact than *Saturn*, for it must give place and preheminence to the other Metals, leaving the victory with them; for it must vanish and be quite consumed with the unfixt inconstant Metals; in it all the three properties of the three principles are most course; and because its Salt is very fluxible above that of other Metals and Planets, so is its Body more fluxible, inconstant, unfixt, and volatile, than any other Metallick Body. As *Saturn* steps to its regeneration, so know that in like manner, as common Water is forced by the natural coldness, by the change of the Heavens, whereby it becomes a coagulated Ice, in like manner is it to be made evident, that by reason of the great coldness which is found to be in the Salt of *Saturn* above other Salts; *Saturn* is also coagulated and made corporal; Ice dissolves into water by heat; so likewise the coagulated *Saturn* is made fluxible by Fire, it hath most of *Mercury* in it, but it is inconstant and volatile; it hath least of Sulphur, and therefore according to its small quantity its cold body cannot be made warm; it hath little Salt, but fluxible, otherwise Iron would be more fluxible and malleable than Lead, if the Salt alone could cause a malleableness and fluxibleness, because Iron contains more Salt than any other Metal: Seeing then there is a difference to be found in this point, you must therefore observe and remember the difference, and how to distinguish between Metals.

All Philosophers have wrote as well as I, that the Salt gives the Coagulation and Body to every Metal; and it is true; but to prove it by an example, how and after what manner this Relation is to be understood: Plume Allom is esteemed to be only a meer Salt, and is approved to be such, which in this particular may be compared to Iron, that the Salt of the Plume Allom is found to be a thing unfluxible as Iron is. On the other side, *Vitriol* likewise is a Salt, manifesting it self in a small quantity, but fluxible and open, therefore its Salt cannot yield such a hard congelation unto its appropriated Metal, as the other can; although all the Salts of Metals grew out of one certain Root, and out of one Seed, yet nevertheless you must observe a difference in their three Principles, as also you must observe & remember, that a difference is found in one Herb from the other, and likewise how man differs from other Creatures and Animals in Qualities, Original, and the three Principles; for one Herb is indued with more of this, another with more of that kind, which in like manner is to be understood concerning Man and other Animals. The Soul of Lead consists in a sweet quality, as also doth the Soul of Tin, and sweeter yet, that nothing almost may be compared to it, being first of all purified to the highest by separation, that the pure be well separated from the impure, that a perfect accomplishment may succeed in the Operation: Otherwise the Spirit of Lead is by nature cold and dry, wherefore I advise, that it be not much used by Men and Women, because it over cools Nature, so that the Seed of both cannot perform their Natural Function; nor doth it much good to the Spleen and Bladder, but in other cases it attracts flegmatick Humours unto it, which raise up much Melancholy in Men; for *Saturn* is a Ruler, and such a *Melancholicus*, whereby a Man is confirm'd in his Melancholy, wherefore its Spirit is used, for one Melancholy Spirit attracts another unto it, whereby Mans Body is freed and delivered from its infused Melancholy. Externally the Soul of *Saturn* is so healing, in all Sores old or new, Cuts, Thrusts, or Accidents by Means or Nature, so that no Metal can do the like; it is cooling in all hot, tumified Members; but Noble *Venus* hath the pre-eminence to mundifie and cauterize all putrid Sores, and to lay a ground for their Cure, which have their access from within; for in her essence she is hot to dry up, but *Saturn* on the contrary is found to be cold in his Essence.

The Celestial Light of the Sun is much hotter than the Light of the Moon; for the Moon is much lesser than the Sun, and according to its dimension and division it contains an eighth part of the greatness in its Circle; if then the Moon in this her Magnitude of the eighth part could excel the Sun, as the Sun excels the Moon, all Fruits and Productions of the Earth must perish, and there would be a perpetual Winter, no Summer to be found at any time: But the Eternal Creator hath in this case well ordained a certain Order and Law for his Creatures, that the Sun should give light by day, and the Moon by night, and so all Creatures should be served. Those Children which are subject to the influence of *Saturn*, are melancholy, churlish, continually murmuring, as old covetous people, who do no good to their own Bodies, and yet never have enough; they put their Bodies to much labour, torment themselves with thoughts and whimsies, seldom recreate themselves, or are merry with other people, nor do they greatly regard the natural love of fair Women.

In brief, I tell thee that *Saturn* is generated of little Sulphur, little Salt, and much unripe gross *Mercury*, which *Mercury* is to be esteemed as a Froth that floates upon the Water, in comparison of that *Mercury* which is found in *Sol*; and is much more hot in its degree, and therefore the *Mercury* of *Saturn* by reason of its great coldness, hath not so quick a running Life as that which is made of Gold, wherein more heat is to be found, whence that running Life hath its original: Therefore in the inferiour world we must take notice of little *Vulcan* in the augmentation and transmutation of Metals, as I have described those three Principles of *Saturn*, as concerning their descent, nature, and complection. And every one must know, that no transmutation of any Metal can follow out of *Saturn*, by reason of its great coldness, only and except to coagulate common *Mercury*; for the cold Sulphur of Lead can qualifie and take away the hot running Spirit of the Quicksilver, if the process be rightly ordered, wherefore it is not amiss to observe, that *Mercury* is so detained, that the Theory should agree with the Practick, and meet together in a certain measure and concordance. You must not therefore quite reject *Saturn*, nor in all points scornfully neglect him, because its Natures and Virtues are known yet but unto few; for the Stone of the wise hath the first beginning of its Celestial, high-shining Colour only out of this Metal, and from the influence of this Planet, the Key of Constancy is delivered unto him by putrefaction, because the red cannot be made out of the yellow, except before-hand a white be made out of the beginning of the black.

I could yet treat variously, and at large of many wonderful works of Natural and Supernatural things. But because other Labours prevent me therein, of making a longer Narration, I therefore put a Conclusion to this Treatise at present, referring the other concerning the concealed Secrets of Minerals until I have a purpose to write further, in a particular Treatise of *Antimony, Vitriol, Brimstone, Magnet,* and which in especial are endowed before others, and depend upon those, out of which Gold and Silver have their beginning, middle and end, together with the true transmutation particularly; which virtues and power they have received out of one thing, wherein all these lie to be generated invisibly concealed, together with all Metals; which matter is publick before the eyes of all men, but because the vertues and powers are very deeply buried and unknown to the most part, therefore this matter is likewise esteemed as nothing, or of no value, and unprofitable, out of ignorance; even as the Disciples of the Lord going to *Emaus*, their eyes were opened at the breaking of Bread, that they knew wonder above wonder, what the rich Creator hath placed in the vile creature, the name is *Hermes*, who carries a flying Serpent in his Shield, having a Wife whose Name is *Aphrodita*, who can know the Hearts of all men, and yet all is one, and one only thing, one only Essence, which is common in all Places, and known every where, every one grasps it with his hands, and uses it in vile matters, and of small value; he values the vile at a high rate, and that which is high he casts away; it is nothing else but Water and Fire, out of which the Earth is generated by the help of the Air, and is yet preserved. Praise be to the most High for his Gifts: At present enough is revealed what my intent was to shew in this Treatise, and so I depart hence; for in separation all is to be found.

Of Natural and Supernatural Things, and Others

Of the Medicine or Tincture of Antimony, *as well to preserve Mans Body in Health, and to divert all desperate, and incurable Diseases, as also to cure the Leprosie of Metals, to purifie and to transmute them into the best Gold.*

Written by that Noble and Learned Philosopher, Roger Bacon.

Stibium or *Antimony*, as the Philosophers say, is composed of a Noble Mineral Sulphur, which they accounted to be the black secret Lead of the Wise.

The *Arabians* call it *Asmat* or *Azmat*; the Alchymists retain the Name *Antimony*.

Addition. The *Moors* call it *Antimony*, others call it *Alabaster*, or *Tarbason*. By the *Arabians* and *Spaniards* it is called *Alcohol. Avicennæ* c. 7. calls it *Artemed. Alexius* of *Piedmont*, in his seventh Book of Secrets, calls it *Talck*, even as *John Jacob Wecker* renders it in his Books of Secrets; but *Talck* is far different from *Antimony. Pliny*, Book 33. Chap. 6. of *Antimony. Dioscorides* gives a preparation of *Antimony*, Book 5. Chap. 39. They call it also *Stibi, Stimmi, &c.* The *Germans* call it *Spies glass*, or as *George Fabricius* would rather have it, *Spies glantz. Gerlandius* calls it Black *Alcophil, Altofel*, or *Alirnu*, others *Cosmet*, and it is twofold, Masculine and Feminine.

It will lead us to the consideration of higher Mysteries, if we behold and discern that Nature wherein Gold is exalted, even as the *Magi* have found that this Mineral is by God ordained under the Constellation of *Aries*, which is the first Celestial Sign, wherein the Sun takes its Exaltation, though this be not regarded by the Vulgar; yet discreet people will know, and the better observe, that even in this place also the Mysteries and Perpetuity may in part be considered with great benefit, and in part discovered.

But some ignorant and indiscreet people think, that when they had *Antimony*, they would deal well enough with it by Calcination, others by Sublimation, and some by Reverberation, thereby to obtain its great Mystery and perfect Medicine. But I tell you, that here in this place it availes not in the least, either Calcination, Sublimation, or Reverberation, whereby afterwards a perfect extraction can or might be done or effected with profit, to transmute the meaner into a better Metallick virtue; for it is impossible for you.

Be not deluded; some of the Philosophers which have wrote of such things, as *Geber, Albertus Magnus, Rasis, Rupecissa, Aristotle*, and many others: But observe this: Some say, that if *Antimony* be made to a *Vitrum* or Glass, the bad volatile Sulphur is gone, and the Oil which may be prepared out of that Glass, will be a very fixt Oil, and will really give an ingress and Medicine of perfection to the imperfect Metals.

These words and opinion are good and true, but it will not be nor appear such indeed; for I tell you truly, without concealed speeches, that if you lose any of the aforesaid *Sulphur* in the Preparation or Burning, for a small fire may easily prejudice it, you then have lost the true penetrating Spirit, which should make the whole Body of *Antimony* to a perfect red Oyl, which should also ascend over the helm with a delightful sent, and curious Colours; observe likewise, that the whole Body of this Mineral, with all its Members, should be but one Oyl, and ascend over the helm without any loss of weight, excepting the *feces*.

How should the Body be brought to an Oil, or yield its pleasant Oil, if it be brought to the last being of its degree, for Glass is in all things the utmost and last.

You shall likewise know that you shall not obtain that perfect noble Oil in the least, if it be extracted with corrected Vinegar poured upon the *Antimony*, nor yet by Reverberation; and although its various colours may appear, yet is it not the right way; you may indeed get an Oil, but you must know that it hath no part of the Tincture, or power of transmutation in it.

Now we come to the Manual Operation.

Take in the Name of God, and of the Eternal Trinity, fine and very pure Mineral *Antimony*, which is fair, white, massie, and inwardly full of yellow Streaks or Veins, and likewise of red and blew Colours, and small Veins, this is the best; pound it to fine Powder, dissolve it by little and little in *Aqua Regis*, that the Water may conquer it. After Solution take it out immediately that the *Aqua Regis* may do it no prejudice; for it will quickly dissolve the Tincture of the *Antimony*; for our Water in its nature is like to the *Ostrich*, which by his heat can digest Iron, and consume it to nothing; for the Water will consume it, and turn it to a Mud, that it shall remain only as a yellow Earth, and then is it quite spoiled.

Take an Example hereof from Silver, which is dissolved, fair, pure and fine in these our Waters, but if it stand a night therein while the Water is strong and full of Spirits, I tell you, your good Silver will be corroded to nothing in these our Waters; and though you would reduce it into a Massie Body, you cannot; for it will remain as a pale yellow Earth, and sometimes it will run together in the form of Horn, or of a white Horse Hoof, which you can by no Art reduce into a Body.

Wherefore you must remember to take the *Antimony* out presently after the Solution, precipitate and adulterate it according to the custom of *Alchymists*, that it may not be corroded with its perfect Oil by the Water, and burnt up to nothing.

The Water wherein we dissolve is thus made.

R. *Vitriol*, a pound and a half, *Salt-Armoniac* one pound, *Azinat* one pound, *Salt-nitre* a pound and a half, *Salt-gemme* one pound, *Allom* half a pound; these are the Ingredients which belong unto the making of the Water for the Solution of *Antimony*.

Take and mix them well together; at first distil very slowly, for the Spirits ascend with greater violence than those of any other common *Aqua fortis*; beware of its Spirits; for their Fumes are very subtile and hurtful in their penetration.

When you have adulterated the *Antimony* well and purely from the corrosive Water, then put it into a clean Vial, poure good distilled Vinegar upon it, set it forty dayes and nights to putrefie in Horse-dung, or in *Balneum Mariæ*, it will be bloud-red. Take it out, and see how much is yet to be dissolved, decant off gently the pure and clear, which is red into a Glass-Gourd, poure other Vinegar upon the *Fæces* as before, that if any thing should yet remain therein, it might be dissolved; this must be done four times in fourty days and nights; for if any good be in the *Fæces*, it will be dissolved in that time, then cast the Dregs away as unprofitable, being but Dirt, and to be cast to the Dunghill.

Put all the Solutions in a glass-Gourd into *Balneum Mariæ*, distil all the tart Vinegar from it, pour it on again, or else pour fresh, if this be too weak, it will quickly dissolve in the Vinegar; distil it again from it, that the Matter be quite dry; then take common distilled water, wash all tartness from it with the Vinegar imparted to the Matter, then dry the Matter in the Sun, which is of a very deep red, or else dry it very well at a gentle fire.

When the Philosophers find our *Antimony* thus secretly prepared, they say then that its external nature and virtue is inverted internally, and the internal cast forth externally, henceforth becoming an Oil, which is concealed in its innermost and profoundest part, till it be well prepared, and cannot any more be brought into its first Essence, untill the last Judgment; and it is true, for so soon as it feels the force of the fire, it flies away in a Vapour with all its parts, because it is volatile.

Some of the common Laborators, having thus prepared *Antimony*, they take one part out because of its consumption, that they may the better operate it, they mix with it one part of *Salt-Armoniac*, one part of the *Vitrum* (with others *Titrum*) one part of the *Rebooth* (with others *Cadoli*) wherewith the Bodies are cleansed; this mixture they cast upon a pure *Luna*, and if there were eight Ounces of the *Luna*, they found ten Drams of good Gold in the separation, and sometimes more; and by this work they gained wherewithal to bear their Charges, the better to attend upon, and attain unto the great Work. The ignorant called this an induction into the Silver, but that is false; for this Gold is not brought into it by the Spirits, but every kind of Silver hath one Ounce of Gold more or less in the Mark (or 8 Ounces) for Gold is so united with the Nature of Silver, that it cannot be separated from it, either by *Aqua fort*, or common *Antimony*, as the *Gold-smiths* know.

But when the aforesaid Composition is cast upon the *Luna* in the flux, then happens such a separation, that the *Luna* doth freely let go the Gold implanted therein into the *Aqua fort*, and is separated from it, letting it precipitate and sink to the bottom, which otherwise could not be done at all. Therefore it is not an induction into the *Luna*, but a bringing out of it.

But we return again to our proposed Work; for we would have only the Oil, which was only known to the Wise, and not to the Ignorant.

When you have rubified the *Antimony* very well according to the former Directions, you must have in readiness a Spirit of wine well rectified, pour it over the red Powder of *Antimony*, set it four daies and nights in a gentle *Balneum Mariæ*, that it may dissolve very well. And if then any of it remain undissolved, pour fresh Spirit of Wine upon it, set it again into the Bath as aforesaid, all will be well dissolved; and if perhaps any more *Fæces* remain, they will be very few, cast them away, for they are good for nothing. Put the Solution into a glass-Gourd, with a Head luted upon it, set it into *Balneum Mariæ*, with its receiver to take the Spirits, distil slowly with a slack heat, till all the Spirit of Wine be come over, pour it in again upon the dry matter, draw it off again as before; this pouring in & abstracting continue so often, till you see the Spirit of Wine ascend over the helm in various colours, then it is time that you follow it with a strong fire, then with the Spirit of Wine ascend red into the helm, and drop into the Receiver like a bloody Oil, and the tender Body ascends like a red Oil, dropping into the Receiver; truly this is the most secret way of the Wise, the so much applauded Oil of *Antimony*; it is a noble, well sented, virtuous, and powerful Oil, as you shall hear afterwards.

But here I will teach and instruct you poor Operators another way, because you have not the Means to attend the great work, not as the Ancients did, with the separation of Gold out of Silver.

Wherefore take one part of the Oil, or half an Ounce of *Saturn*, four Ounces calcined according to Art, pour the Oil upon the *Calx* of *Saturn*, mixing it, set it ten daies and nights in the heat, into the secret Furnace; every two days augment the fire one degree, according to the capacity of the Furnace; after four days and nights set it into the third degree of Fire, therein let it rest three days and nights, then open the Door or Vent of the fourth degree, which must likewise continue three days and nights; afterwards take it out, the *Saturn* will be above black, like unto Charcole dust, but under this black dust you will find other Colours, throughout pure, red, yellow, which flux with *Venetian Borax*, you will find it converted into good

Gold by the power of our Oil, so have you means again to set forward the great work.

We return again to our purpose, where we left off before. You have heard, and have been instructed how to abstract the Spirit of Wine with the Oil over the helm into the Receiver, and to use it for the work to convert *Saturn* into Gold. But we will now hasten to the other work of the Tincture, and give advice concerning it. It will therefore be necessary to separate the Spirit of Wine again from the Oil, which do as followeth;

Take the mixture of the Spirit of Wine, and of the Oil, set it into *Balneum Mariæ*; distil the Spirit of Wine only from the Oil with a very slack heat, so that you may be assured that there is no more of the Spirit to be found in this most precious Oil, which you may easily try; when you see some of the drops ascend over with the Spirit of Wine, it is a sign that the Spirit of Wine is separated from the Oil, then remove all the fire from under the Bath, how little soever it be, that it may cool the sooner. Take away the Receiver with the Spirit of Wine, stop it very close, for it is full of Spirits which it hath retained from the Oil, as you will hear afterwards: But in *Balneum Mariæ* you will find that blessed Oil of *Antimony* red as Bloud; take it out, wash the Lute off by gentle mollification, that nothing impure may fall into that curious red Oil, when you take the head off; reserve it carefully, that by no means it may receive prejudice, for you have a Celestial Oil, which in a dark night shines like a glowing Cole, and this is the reason, because its internal power and soul is cast forth externally, the hidden Soul being now revealed, shining through the pure Body as a Candle through a Lanthorn, even so at the last day, these our invisible internal Souls shall be revealed, and seen out of the Body, shining as the clear Sun: So keep each apart, as well the Spirit of wine full of power, and wonderful in curing humane Distempers, as also the blessed, red, noble, celestial Oil, which transmutes all the Diseases of the imperfect Metals into the perfection of Gold; and the power of the spiritual Wine extends very far being rightly used.

I tell you, you have obtained a Celestial Medicine, to cure all the Diseases and Distempers of Mans Body; its use is, as followeth;

Of Natural and Supernatural Things, and Others

In the Gout.

Give three drops in a Cup of Wine fasting to the Party, just at the time when he feels the beginning of his misery, anguish and pain to come upon him, the second and third, use it in like manner; it allaies all pain the first day how great soever it be, and prevents Swelling; the second day it causes Sweat, which is very nasty, tough and thick, very soure in taste, and of an evil sent, and most of all in those parts where the Members are united and joined together by the Joints; and if you should give none in the third day, yet will there be a purgation of the Veins, and of the Excrements, without any molestation or pain; is not this a great power of Nature?

In the Leprosie.

At the first time take six drops fasting, and cause the impure party to be alone, free from sound people, in a place far distant, and commodious; for all his Body will begin to send forth Fumes and Steams, like unto a stinking Fog, and Vapours abundantly; the next will Scales and much Uncleanness fall from his Body; then let him have three drops of this Medicine, and let him take it in on the fourth day, afterwards on the eighth or ninth day by the assistance of Gods Grace and Blessing, he will be quite clean.

In the Apoplexie.

Let one drop fall upon the tongue of the Patient, it will attract it forth immediately like unto a Mist or Fume, and restore the party again; but if he were taken in the Body, or in the Members and Limbs, then give him three drops at once in good wine, as you have been taught in the Gout.

In the Dropsie.

Give one drop in baulm water, or Valerian water six days together, the seventh day give three drops in good wine, and it is sufficient.

Of Natural and Supernatural Things, and Others

In the Falling Sickness, and its kinds, as Epilepsie, Catalepsie, and Analepsie.

In the beginning of the Fit give the Patient two drops in Sage-water, after three hours, give him three drops more, and it is sufficient. But if in case any thing should stir again, give him two drops, as hath been said.

In a Hectick.

Give the party two drops the first day in water of Violets, the second day two drops more in good Wine.

In Agues.

Give the party three drops in the beginning of the Fit, early in a morning, in good distilled water of St. *John's* wort, or of Succory, and the next day two drops more fasting.

In the Plague.

Give the Patient seven drops in good Wine, let the infected party be alone, and let him sweat well upon it, by the Divine Assistance that poison will not prejudice him as to his Life.

For a prolongation of a healthful Life.

Take and give two drops at the beginning and entrance of the Spring, and in the beginning or entrance of Autumn likewise two drops; every one that so takes it, is freed, and well preserved from unhealthful and infectious Air, except the Disease were by Almighty God ordained for the death of the party.

We will now step further to the Oil, and its Power, and shew how by it the Diseases of the impure Bodies of the Metals may be cured.

In the Name of God, take very pure, fine, refined Gold, as much as you will, or think to be sufficient, dissolve it in a rectified Wine, as is usual to make *Aqua vitæ*; after solution of the Gold, set it a Moneth in digestion; this distil in a Bath very slow and gently, distil the Spirit of Wine divers times from it, so long till you see your Gold lie at the bottom like a Juice: This is the true way and meaning of some of the Ancients, to prepare Gold. But I will shew and teach you a way much readier, better, and more beneficial; that in stead of this prepared Gold, you take one part of the *Mercury* of Gold, as I have taught the making of it in another place; abstract from it its water of Airiness, that it may be a subtle Dust, and take two parts of our blessed Oil, poure the Oil very slowly upon the Dust of the *Mercury* of Gold, till all be in it, set it in a Vial well sealed, in the heat of the first degree of the secret Furnace; therein let it stand ten dayes and nights, your Powder and Oil will be quite dry, of a black gray colour. After ten days give it the heat of the second degree, the gray and black colour will by little and little become white, till at last it will be of a heavenly white, and at the end of the ten days it will begin to be of a pure red, but let not this trouble you; for all these Colours proceed only from the *Mercury* of Gold, which swallowed up our blessed Oil, and now conceals in the innermost part of its Body; but our Oil will conquer this *Mercury* of Gold by the power of the fire, and cast it forth from within, and the Oil will predominate over it with its hot red Colour, and be continually outwards. And therefore it will be time, after the expiration of twenty days, that you open the window of the third degree, wherein the external white Colour and Power will by little and little enter in into the inward part, and the internal red Colour will turn outward by the force of the fire. Keep this degree of heat ten days without diminution or augmentation of it, you will see a Powder which before was white, to be now very red, but let not redness trouble you, for 'tis yet unfix'd and volatile. And after these ten days are ended, thirty days being in all expired, then open the last window of the last degree of fire, keep it ten days in this degree, this high red pouder will then begin to flux, let it stand so in flux the ten days, then take it out, you will find at the bottom a very high, red, transparent stone of a Ruby Colour, flux'd according to the form of the Glass, as is taught in the Treatise of *Vitriol*, wherewith you may make projection. Praise God for such his high Revelation, and thank him for ever, *Amen.*

Of Natural and Supernatural Things, and Others

Its Multiplication.

The Ancient Wise, having found the Stone, and prepared it to a perfect power, and mutation of the imperfect Metals into Gold, have a long time enquired whether a thing were not to be found to augment the power of the Stone; and they found two kinds of Augmentation, one of the power of it, so that the Stone may be brought much higher; of this multiplication you will find direction in the Treatise of Gold. The other Augmentation is an augmentation of the quantity of the Stone, in its former power, so that it receives no more, nor loses any thing of its power, though it increase in weight, and augment more and more, that out of one Ounce many Ounces arise and increase.

The Augmentation or Multiplication is done as followeth; Take your Stone in Gods Name, grind it to a subtile powder, add to it as much of the *Mercury* of Gold, as is taught before, put them together into a fine round Vial, seal it hermetically, set it into the fiery Furnace, proceeding as you have been instructed before, only this time is shorter, for whereas before you had ten (thirty) days, now you need no more than four (ten) days, otherwise the work is one and the same.

Praise and give thanks to Almighty God for his high Revelation, continue in Prayer for his Grace and Divine Blessing in this Art and Operation, as likewise for continuance of Health and Prosperity; withal let the poor be recommended to your Help and Charity.

Glory be to Almighty God

Of Natural and Supernatural Things, and Others

A Work of Saturn, *of Mr.* John Isaac Holland.

The PREFACE.

Courteous Reader,

The PHILOSOPHERS *have written much of their Lead which is prepared out of* Antimony, *as* Basilius *hath taught; and I am of the opinion, that this Saturnine Work of the most excellent Philosopher M.* John Isaac Holland *is not to be understood of common Lead, (if the Matter of the Stone be not much more thereby intended) but of the* Philosophers *Lead. But whether the Vulgar* Saturn *be the Matter of the Philosophers Stone, thereof you will receive sufficient satisfaction from the subsequent 17 Considerations or Documents. This is published for the benefit of all the Lovers of this Art, because it expounds and declares the Stone of Fire.* Vale.

A Work of Saturn

In the Name of the Lord, Amen.

My Child shall know, that the Stone called the *Philosophers Stone*, comes out of *Saturn*. And therefore when it is perfected, it makes projection, as well in mans Body from all Diseases, which may assault them either within or without, be they what they will, or called by what name soever, as also in the imperfect Metals.

And know, my Child, for a Truth, that in the whole vegetable work there is no higher nor greater Secret than in *Saturn*; for we do not find that perfection in Gold which is in *Saturn*; for internally it is good Gold, herein all Philosophers agree, and it wants nothing else, but that first you remove what is superfluous in it, that is, its impurity, and make it clean, and then that you turn its inside outwards, which is its redness, then will it be good Gold; for Gold cannot be made so easily, as you can of *Saturn*, for *Saturn* is easily dissolved and congealed, and its *Mercury* may be easily extracted, and this *Mercury* which is extracted from *Saturn*, being purified and sublimed, as *Mercury* is usually sublimed, I tell thee, my Child, that the same *Mercury* is as good as the *Mercury* which is extracted out of Gold, in all operations; for if *Saturn* be Gold internally, as in truth it is, then must its *Mercury* be as good as the *Mercury* of Gold, therefore I tell you, that *Saturn* is better in our work than Gold; for if you should extract the *Mercury* out of Gold, it would require a years space to open the body of Gold, before you can extract the *Mercury* out of the Gold, and you may extract the *Mercury* out of *Saturn* in 14 days, both being alike good.

Would you make a work out of Gold alone, you must labour two whole years upon it, if it shall be well done: and you may finish a work of *Saturn* in 30 or 32 weeks at the most. And being both well made, they are both alike good; *Saturn* costs nothing or very little, it requires a short time, and small labour; this I tell you in truth.

My Child, lock this up in thy heart and understanding, this ♄ is the Stone which the Philosophers will not name, whose name is concealed unto this day; for if its name were known, then many would operate, and the Art would be common, because this work is short, and without charge, a small and mean work.

Therefore doth the name remain concealed; for the evils sake which might thence proceed. All the strange Parables which the Philosophers have spoken mystically, of a Stone, a Moon, a Furnace, a Vessel, all this is *Saturn*; for you must not put any strange thing unto it, only what comes from it, therefore there, is none so poor in this world, which cannot operate and promote this work; for *Luna* may be easily made of *Saturn*, in a short time, and in a little longer time *Sol* may be made out of it. And though a man be poor, yet may he very well attain unto it, and may be employed to make the *Philosophers Stone*.

Wherefore my Child, all is concealed in *Saturn*, which we have need of, for in it is a perfect *Mercury*, in it are all the Colours of the world, which may be discovered in it; in it are the true black, white and red Colours, in it is the weight, it is our *Lattin*.

Example.

The eye of a man cannot endure any thing that is imperfect, how little soever it be, though it be the least Atome of Dust, it would cause much pain, that he can rest no where. But if you take the quantity of a Bean of *Saturn*, shave it smooth and round, put it into the Eye, it will cause no pain at all; the reason is, because it is internally perfect, even as Gold and Precious Stones. By these and other Speeches you may observe, that *Saturn* is our *Philosophers Stone*, and our *Latten*, out of which our *Mercury* and our Stone is extracted with small Labour, little Art and Expence, and in a short time.

Wherefore I admonish you, my Child, and all those who know its name, that you conceal it from people, by reason of the evil which might thence arise; and you shall call the Stone our *Laton*, and call the Vinegar Water, wherein our Stone is to be wash'd; this is the Stone and the Water whereof the Philosophers have wrote so many great Volumes.

There are many and different works in the Mineral Stone, and especially in that Stone which God hath given us *gratis*, whereof many strange Parables are written in the Mineral Book.

But this is the true Stone, which the Philosophers have sought, because it makes projection upon all the imperfect Metals, especially upon quick *Mercury*, and moreover it makes projection upon all diseases whatsoever, which may come into mans Body, as likewise upon all Wounds, *Cancer*, *Fistulaes*, *open Sores*, *Buboes*, *Imposthumes*, and all whatsoever can come externally upon mans Body, therefore this Stone is not under the Mineral work, but under the Vegetable.

It is the beginning of the Vegetable Book, and the principal; this Stone is called *Lapis Philosophorum*, the Mineral Stone is called *Lapis Mineralis*, and the third Stone is called *Lapis Animalis*. This Stone is the true *Aurum potabile*, the true Quinessence which we seek, and no other thing else in this world but this Stone. Therefore the Philosophers say, whosoever knows our Stone, and can prepare it, needs no more, wherefore they sought this thing and no other.

My Child shall take 10, 12, or 15 pound of *Saturn*, wherein is no mixture of any other Metal; laminate it thin, have in readiness a great Stone Jugg, half full of Vinegar, stop the Jugg very close, set it in a

Lukewarm Bath, every three or four days scrape off the calcin'd *Saturn* from the Plates, and reserve it apart, thus do so long till you have 5 or 6 *l.* of the calcin'd *Saturn*, then grind it very well on a Stone with good distilled Wine-Vinegar, so as you may paint therewith, then take two or three great Stone-pots, therein put the *Calx* of *Saturn* which you ground, poure good distilled Wine-Vinegar upon it, that two parts of the Pot be full, stir it well together, stop the Pot close with a polished Glass or Pebble-stone, set the Pots in a Bath, stir it four or five times in a day with a wooden Ladle, lay the Glass or Stone Stopple again over it, make the Bath no hotter than that you may well endure your hand therein, that is, lukewarm; so let it stand fourteen days and nights, then decant that which is clear into another Stone-pot, poure other distilled Vinegar upon the *Calx* which is not well dissolved, mix them well together, set it 14 days in the Bath, again decant it, and poure other Vinegar upon it as before. This decantation and pouring on continue so long till all the *Calx* of *Saturn* be dissolved, then take all the dissolved *Saturn*, set it in a Bath, evaporate the Vinegar by a small fire, the *Saturn* will become a powder or lump. Or stir it about until it be dry, you have a mass or powder of a dark yellow, or honey colour, then grind the powder again very finely upon a Stone with distilled Vinegar; put it into a stone-pot, stir and mix it well together, set it again into a Bath, which is but lukewarm so let it stand five or six dayes, stir it every day from the top to the bottom with a wooden Ladle, cover it again with the glass-Stopple, then let it cool, poure off that which is dissolved into a great stone pot, poure other Vinegar upon it, mix and stir them well together, set it into the Bath as before, reiterate this decantation and pouring on so often, till no more will dissolve, which try with your tongue, if it be sweet, it is not enough dissolved, or put some of it into a glass-gourd, let it evaporate, if any thing remain, it is not yet all dissolved which would be Gold, and then what remaines in the pot are *Fæces*, and sweet upon the Tongue; if you find any thing in the Gourd, it is not yet all dissolved, then may you poure fresh Vinegar upon it, till all be dissolved, then coagulate it as before, poure other Vinegar upon it, stir it, set it again into the Bath, reiterate this operation of solution and coagulation so long till you find no more *Fæces* at the bottom, but all be dissolved into a pure clear water, then is *Saturn* freed from all its Leprousness, Melancholy,

Of Natural and Supernatural Things, and Others

Fæces, and blackness, being pure and white as Snow, for it is cleansed from all its uncleanness, because its coldness stands outwards as *Luna* doth, and its heat is internal, fluxible as wax, and sweet as sugar Candy.

Why is it as white as Snow?

Because it is purified from all its impurities, and because its coldness stands external as *Luna* doth, and its heat is internal.

Why is it sweet?

Because the four Elements in it are pure, and separated from all sulphurous stink and blackness, which *Saturn* received in the Mine; it is almost Medicinal, and like unto Nature: And because it is so pure, it affords some of its internal virtue outwardly, as that of Sweetness; but the heat is so covered with the cold, that it cannot put forth its power externally by reason of the cold which is external (the heat of *Saturn* lies internal, even as in *Salt-Nitre*) as doth the Taste, the Spirit of Tasting is the most subtile in all things, as is taught more at large in the Book of Vegetables, how the Air doth dilate it self from all Herbs and Flowers externally; for the Spirit of the Air lies in the inward part of all things; for God created nothing in this world but it hath its peculiar Taste or Air, the Air and the Taste are one Spirit, the Taste goes out of the Air, as Smoke from the Fire.

But how comes it to pass, that a thing which hath a sweet Air, is bitter in Taste? The cause is, because the *Fæces* of that thing are putrid and stinking in the Elements, that is the Choler or Heat; for whatsoever is unnaturally hot, hath a bitter Taste; the Air and the Taste are both one Spirit, and as the Spirit of the Air presses outwards through a hot thing, so doth the Air embrace the Taste about, and descends the subtile Taste, that it should not be burnt by the vehement burning Choler, as in the Herbal is at large express'd.

But the cause why *Saturn* is sweet in Taste is, that it is almost pure and clean, having scarce any unnatural heat in it, which can burn the subtile Tast, therefore it hath the Taste externally, and the Taste hath the Spirit of the Air lock'd up in it.

Of Natural and Supernatural Things, and Others

My Child, know what I said before, that a thing wherein is much burning heat, the Air locks up the Taste therein, because the Taste shall not be corrupted by the unnatural heat. So the Taste includes the Air in it, when it issues forth from a thing which is externally cold; for the subtil Spirits of the Air or Sent of a thing can endure no Cold, as we see daily in Herbs and Flowers that they yield no Sent in the Winter, as they do in the Summer; but they hide themselves in the Winter, and the Spirit hath the Sent inclosed in it, and the Spirit of Sent or Air. Behold a man that hath taken Cold, immediately he loses his Sent, and his Tasting is diminished. Even so it is here with *Saturn*; it is quite cold, so that the Taste manifests it self with the Spirit of Sent; for the Spirit of the Taste hath the Smell in it. Look upon Sugar which is well clarified from its *Fæces*, how sweet it is in Taste, yet it yields no Sent, yet there is an extraordinary sweetness in Sugar. What is the reason of this? Sugar is very cold externally, therefore is it white as Snow, and of a sweet Taste; yet Sugar internally is hot and moist, of the temper of Gold, and of such great virtue that it is called the Philosophers Stone, as it is approved, and very prevalent to cure all the Distempers of mans Body, as appears by its operation. The reason why I say this, my Child, is, that you should altogether understand its internal & external, and the Spirits which are in these things, whereof we discourse; that thereby you should know Gods wonderful works, and what wonders he works in these inferiour things, which are all made for our use.

What hath God in us, for whose sake he hath created all these Wonders, and all these things?

Wherefore, my Child, believe in God, love him, and follow him, for he loves you, as he makes it appear, and manifests himself in all things, as well in their Internals as in their Externals. O how wonderful is our Lord and God, from whom all Wonders proceed!

Now, my Child, why is **Saturn** *fluxible as Wax?*

By reason of its abounding *Sulphur*, which is therein; for I find no fluxibleness or fusibleness in any thing saving in *Sulphur*, *Mercury* and *Arsenick*, and all these three are in *Saturn*; so that *Saturn* is quickly fluxible, but all these three are cleansed with it from their uncleanness. And do you not know, that the Philosophers call their Stone *Arsenick*, and a white thing; and they say their *Sulphur* is incombustible; they call it likewise a red thing, all this is *Saturn*, in it is *Arsenick*; for *Luna* is principally generated of a white *Sulphur*, as is plainly taught in the Book of *Sulphurs*, and all *Arsenick* is internally red as Bloud, if its inward part be brought outwards, as is demonstrated in the Book of Colours, *&c*. *Saturn* stands almost in the degree of fix'd *Luna*. So that in it there is a red Sulphur, as you see, when its internal is placed outwards, it will be red as a Ruby; there are no Colours but in the Spirits, so that there is in it a red and a yellow Sulphur. In it is *Mercury*, as may be seen, for *Mercury* is extracted out of *Saturn* in a short time, and with little labour.

So that all three are in *Saturn*, but they are not fix'd therein, but they are clean, pure, incombustible, fluxible as Wax; in it are all things which the Philosophers have mentioned. They say, our Stone is made of a stinking menstruous thing: What think you, is not *Saturn* digg'd out of a stinking Earth? for divers are killed with the ill Sents and Vapours where *Saturn* is digg'd, or they live not long who labour in that stinking black Mine, whence *Saturn* is digg'd. And the Philosophers say, our Stone is of little value, being unprepared; they say, the poor have it as well as the rich, and they say true; for there are not poorer or more miserable people to be found than those which dig and work *Saturn* in the Mine; and they say it is to be found in all Towns and places, wheresoever you come *Saturn* is there. They say it is a black thing: What think you, is it not black? They say, it is a dry water, if Gold or *Luna* be to be refined upon the test, must it not be done with *Saturn*? they must be wash'd and tried with it, as a foul garment is made clean with Sope. They say, in our Stone are the four Elements, and they say true; for the four Elements may be separated out of *Saturn*. They say, our Stone consists of Soul, Spirit and Body, and these three become one. They say true; when it is made fix'd for the white *Mercury* and Sulphur with its Earth, then these three are one.

Whereby is to be observed, that the Philosophers have said true; they concealed its Name for the ignorants sake, who are not their Children, to keep them still in their Ignorance. Thus, my Child, the Ancients took care to conceal the name of the Stone; now let us return to our purpose.

You have now *Saturn* wash'd and cleansed from all its impurity, and made as white as Snow, fusible as Wax, but is it not fix'd yet; we will make it fix the *Mercury* and Sulphur with its Earth.

Take a Glass-Vial, put half of your purified *Saturn* into it, reserve the other half till you have occasion to use it; lay a polish'd Glass upon the mouth of the Glass, set it in a Cuple with sifted Ashes upon a Furnace; or set it on the *Tripos* of Secrets, or in the Furnace wherein you calcine Spirits; give it Fire so hot as the heat of the Sun at *Midsummer*, and no hotter, either a very little hotter, or a very little cooler, as you can best hit it. But if you give it a greater heat, such as you may keep Lead in flux, then your Matter would melt as if it were Oil; and having stood so, ten or twelve days, its Sulphur would fly away, and your Matter would all be spoiled, for the Sulphur which is in your Matter is not yet fix'd, but is in the external. Wherefore the Matter melts presently, and though it be clean, yet it is not fix'd; wherefore give so gentle a fire to it, that it may not flux; so keep it six weeks, then take out a little of it, lay it on a glowing hot Plate, if it immediately melts and fumes, it is not yet fixed, but if the Matter remain unmelted, the Sulphur is then fix'd which is therein; then strengthen the Fire notably, till the Matter in the Glass begins to look yellow, and continually more and more yellow, like to powdered Saffron, then augment the fire yet stronger, till the Matter begin to be red, then prosecute your Fire from one degree to another, even as the Powder becomes redder and redder by degrees, so hold on your Fire, till the Matter be red as a Ruby, then augment the Fire yet more, that the Matter may be glowing hot, then is it fixt, and ready to pour the curious Water of Paradise upon it.

My Child must know, that there are two ways of pouring on the Water of Paradise; I will teach you to make and prepare both, then may you take which you will; for the one is half as good again as the other.

Of Natural and Supernatural Things, and Others

My Child, you may remember, that I ordered you to reserve the one half of the purified *Saturn*, which take and put into a Stone-pot, pour upon it a bottle or more of distilled Wine-Vinegar, set a head on, distil the Vinegar again from it in a Bath, the head must have a hole at the top to pour fresh Vinegar upon the Matter, and abstract the Vinegar again from it, pour fresh Vinegar again on, and again abstract it, this pouring on, and abstracting or distilling off must continue so long, till the Vinegar be drawn off as strong as it was when it was put in, then is it enough, and the Matter hath in it as much of the Spirit of Vinegar as it can contain; then take the Pot out of the Bath, take off the head, and take the Matter out, and put it into a thick glass which can endure the Fire, set a head on it, put it in a Cuple with Ashes, which set on a Furnace, first make a small Fire, and so continually a little stronger, till your Matter come over as red as Bloud, thick as Oil, and sweet as Sugar, with a Celestial Sent, then keep it in that heat so long as it distils, and when it begins to slack, then increase your Fire till the Glass begin to glow; continu this heat till no more will distil, then let it cool of it self, take the Receiver off, stop it very close with Wax, take the Matter out of the Glass, beat it to powder in an Iron Mortar, with a steel Pestle; and then grind it on a Stone with good distilled Vinegar, put this Matter so ground into a Pot, poure good distilled Vinegar upon it, that two parts be full, set the Pot into a Bath with a head upon it, distil the Vinegar off, poure fresh Vinegar again upon it, distil it off again: thus do so long, that the Vinegar be as strong as it was when it was first poured upon it, then let it cool, take the Matter out of the Bath, take the head off, take the Matter out of the Pot, put it into a stronger round Glass which can endure the Fire, as you did before, set it upon a Furnace in a Cuple with sifted Ashes, set a head on, and a Receiver luted to it, then distil it, first with a small fire, which augment by degrees, till a Matter come over red as Bloud, and thick as Oyl, as aforesaid; give it fire till no more will distil, then let it cool of it self, take off the head, break the glass-pot, and take the Matter out, powder it again, and grind it on a Stone with distilled Vinegar, put it again into the Stone pot, poure fresh Vinegar upon it, set it into the Bath, and its head on, distil the Vinegar from it, poure it on again as hath been taught, till the Vinegar remain strong as it was.

Reiterate this distillation in the Bath until the Matter hath no more Spirit of the Vinegar in it, then take it out, set it in a glass-pot, distil all that will distil forth in ashes, till the Matter become a red Oil, then have you the most noble water of Paradise, to pour upon all fix'd stones, to perfect the Stone; this is one way. This water of Paradise thus distilled, the Ancients called their sharp clear Vinegar, for they conceal its name.

My Child, I will now teach you other ways to make the Water of Paradise; this is an easie way, but not so good, nor doth it that high projection in humane Medicines, yet it cures all Diseases within and without, but the other cures miraculously in a short time.

The second way of preparing the Water of Paradise.

My Child, if you would make it after this manner, you must take the half of your prepared *Saturn* which I ordered you to keep, upon which poure the half of your fix'd and prepared Water of Paradise, take the half, put it into a Stone-pot, poure weak Wine Vinegar upon it, mix it well together, then take two pounds of calcined *Tartar*, which is well clarified by solution and coagulation, so that it leave no more *Fæces* behind it, *Salt Armoniac* one pound, which is likewise so clearly sublimed, that no *Fæces* remain after its sublimation, pound both together to a Powder, put them speedily into a pot, and stop it close immediately, or else it will run out; for so soon as the *Tartar* and *Salt Armoniac* come to the Vinegar, they lift themselves up, and would immediately run out of the mouth of the pot, wherefore stop the pot presently, set the pot in a Vessel of Water, they will cool speedily, otherwise if the cold and hot Matter should come together suddenly, they would contest together, rise up, and become so hot, that the pot would break for heat, if it were not set in cold Water; therefore take heed, when you put the powders in, that you stop it immediately, and set it in cold Water before you put the other Powder to it, then will they unite, let them stand a day and a night in that Vessel, then take them out, set them into a lukewarm Bath two days and nights, let it cool of it self, take the Stopple off from the pot, and set a head on, set the pot in sifted Ashes, upon a Furnace, distil with a small fire, and continually greater till all the Vinegar be over, then augment your Fire notably, till you see quick *Mercury* drop out of the Pipe, when it ceases to drop, then augment the Fire by little and little and drive it so long as it drops; you may observe when it will leave dropping, if in the space of one or two *Pater-nosters* one drop doth fall, then augment the Fire till the pot glow at the bottom, for twelve hours and when the *Mercury* is over, then should the *Salt Armoniac* sublime up into the head, and the *Tartar* remain with the Body of *Saturn* at the bottom of the Pot, which take out, put it into a Linnen Bag, hang it in a moist Cellar, the *Tartar* will dissolve, receive it in a Glass, the body of *Saturn* remains in the Bag, take it out, and calcine it in a reverberating Furnace three days and nights, with a great heat, as is taught elsewhere, then extract the Salt out, as is taught in the Mineral Book. You may make projection with the Salt, and coagulate your *Tartar* again, it will be as good or better than it was, likewise take your *Salt Armoniac* out of the Head, it is good again, and if you could have no *Salt Armoniac*, then take three pound of calcined *Tartar*, likewise so clarified, that it leave no *Fæces* behind, you then need no *Salt Armoniac*, therewith may you likewise extract the *Mercury* out of *Luna* and *Jupiter*, wherewith you may do wonders, as is taught in the Miner. Book, where is spoken of the Quintessence of Metals.

Now my Child must know, that this *Mercury* or Quintessence of *Saturn* is as good in all works as the *Mercury* of *Sol*, they are both alike good, and herein all Philosophers agree. My Child, take this *Mercury* of *Saturn*, so drawn out of the Receiver, put it into a Glass Box.

I have now taught you to make two sorts of the Water of Paradise; and know, my Child, that the first way is the best; though it be made with some danger, longer time, and more charge; for the Vinegar is all good, yet the red Oil is the best; its time is alike unto the end, and though it be more tedious before you obtain the red Oil, yet it fixes it self in a short time, if it come to the Matter or fix'd Stone, into a simple Essence in greater redness; but when the *Mercury* comes to the fix'd stone, it holds on a long time in ascending and descending before it die, and when it is quite dead, it makes the red fix'd Stone again into a fixt colour, so covering the red stone with its coldness, that the red stone becomes white again, then must you boil it again gently with a small Fire, till it begin to be yellow, prosecuting the

Of Natural and Supernatural Things, and Others

Fire from one degree to another, as the Colour is higher and stronger, and that so long till it attain to a perfect redness, which requires a long time before it be done, which is not requisite in the red Oil; for the red Oil dies or coagulates forthwith the stone, the one fixing it self with the other into a simple Essence, in a short time. Therefore I tell thee, my Child, that the time of the Oyl is alike long in the end, though it appear to be of a shorter time with the *Mercury*, but it is equally long at the end of the Work, therefore I tell you the Art of both Works, that you may the better understand the Art to make the Oyl from the innermost Nature of the Stone, which is found afterwards.

The Oyl was unknown to the Ancients, for my Grandfather with his Companions found it with great labour and length of time.

So there are two ways to dissolve the Stone, and to poure upon it the clear water of Paradise. Our Ancestors called the Oyl their sharp Vinegar; therefore, my Child, keep the Name private, and I will teach you first of all how you shall join the *Mercury* to your Stone, which you extracted out of *Saturn*, to dissolve it; afterwards I will teach you to bring over the helm that red Oil which you extracted out of your prepared *Saturn*, into a fixt stone, to dissolve your stone.

My Child, weigh your fixt stone, take half as much of your *Mercury*, poure it upon the stone in the Glass, cover the Glass again with a polish'd Glass which may just fit it, set it in a Cuple with sifted Ashes, make a small Fire like the Suns heat at *Midsummer*, and give no more Fire to it, until the Water of Paradise or *Mercury* become all a dead Powder. And know, my Child, that the red or fixt Stone, which before was darkned, when it hath drunk up the Water of Paradise, or *Mercury*, or how you will call it, that it be a Powder between black and gray, then augment the Fire from one degree to another, till the Matter be perfect white, and when it is white, strengthen the Fire yet more, from one degree to another, till it be of a dark yellow Colour, then make it yet stronger, till it be of a perfect red; then rejoice, for your Stone is perfect, and fluxible as Wax. Praise God, who gives unto us part of his Miracles; and do good to the poor; you may see it with your fleshly Eyes, and use Gods goodness miraculously in this corrupt Life, for I tell you in good Charity, that if any one principally attain to this Stone, that it is given, afforded, and lent him from God. Whosoever hath this Stone, may live in a healthful state, to the last term of his Life, appointed him by God, and may have all whatsoever he desires on Earth.

He shall be loved and esteemed of all people, for he can cure them all internally and externally of all Diseases which may befall them; but if the Stone doth not so, it is false, and deserves not the name of the Vegetable Stone, or Philosophers Stone.

Therefore my Child, if God give you this Stone, look diligently to it, that you keep your self from offending God, that you make not this Stone on earth to be your Heaven; govern and rule your self to Gods glory and to the comfort of poor people, that Gods praise may be augmented, to the defence of the Christian Religion, and to the relief of poor exiled Christians I tell you, my Child, if you use it otherwise, God will leave you here a little while to your own Will, but afterwards he will speedily send a punishment, either you shall be struck dead, or die by a Fall; or die some other sudden death, and go Body and Soul to Hell, and be damned eternally, for your Ingratitude to God, who so graciously vouchsafed you so precious and great a Gift.

Therefore, my Child, look carefully to it, so to govern your self to Gods Glory, and the Salvation of your Soul, that the eternal Curse may not fall upon you; and therefore I have left you this Writing as my Testament. Enough hath been said to the wise, therefore look to your self.

The Multiplication of the Stone now perfected.

Now my Child, you may take the half of your Powder, put it into a Glass and melt it, have in readiness a Mould made hollow, of Box-wood, great or small as you please, it must be made smooth and even within with an Instrument, anoint it with Oil Olive, and when your red Powder is flux'd, poure it into the Mould, it will be a precious Stone, red as a Ruby, clear and transparent, take it out of the Mould, and make projection upon the imperfect Metals, and in the Body of Man.

Take ten times as much of prepared *Saturn* as I taught you before, by Coagulation and Solution, till it leave no *Fæces* behind, then take your precious red Powder out of the Glass, that two parts be full, set it into your warm Bath, and let it dissolve: when any thing is dissolved, decant off that which is clear on the top into another Glass, poure other Vinegar upon it, let it dissolve again as before, decant and poure fresh Vinegar upon it so often, till all be dissolved into a clear Water, which is done usually in ten or twelve days, then set all that which is dissolved into a Bath, and a head upon it, distil the Vinegar from it again, and coagulate the Matter so long till it be dry and shine, then put it into another Glass, which set upon a Furnace in a Cuple with sifted Ashes, laying a polish'd Glass upon the Mouth of the Glass.

My Child, know that your Matter is become fixt with the Stone in the solution, make an indifferent hot fire in the furnace, so hot as the heat of the Sun at *Midsummer*, or somewhat hotter; till the Matter begin to be yellow, then go on with the Fire from one degree to another, till you have a perfect yellow, then increase the Fire from one degree to another, till you have a perfect redness, which is quickly done, in half the time for the colour to come, and in the multiplication, but operate as before in the beginning, and poure Paradise water upon the Stone, as was taught you before in this Work, boil and mortifie it in every point to a perfect redness as hath been taught.

Then may you again take half of it out, and make projection therewith, and multiply the other half again in all points as abovesaid, so may you always continue working.

Now I will teach you the other way, and the best that is to water your red fixt Stone or powder with the red Oil, that it be fusible; you must know how much your red powder weighs, then take half the weight of your red Oil, to the full weight of the Stone, and poure it upon the red powder, and when the Oil is poured into the Glass, you may set a small head on, upon a Furnace in sifted Ashes, joining a Receiver to the Nose of the head, make a small fire under it, as the heat of the Sun in *March*, and no hotter; for there is yet some moisture of the Vinegar in the Oil, that it may be abstracted, continue it in that heat, that can perceive no moisture in the Head, then augment the fire a little, as the heat of the Sun at *Midsummer*, and if there be yet more moisture in it, you will perceive it in the head, but if you perceive it not in 6 or 8 days, then take the head off, and lay the polish'd Glass again upon the mouth of your Glass, increase the fire, that you can scarce endure your hand or finger in the Ashes an *Ave-Mary* while, continue the fire in that heat till the red Oil be all fixt with the Powder in the Glass, which you may know thus;

Take a little of the powder out of the Glass, lay it on a glowing Silver Plate, if the powder melts as wax, and penetrates through the Plate as Oil doth through a dry Leather, and makes it Gold throughout, as far as the powder went, then is the Stone finish'd, and if it do not this, you must then let it stand in that heat till it do so without fuming.

Now, my Child, when the Stone is finish'd, take half of it out of the Glass, put it into a Glass melting-pot, and melt the powder gently, which should be done presently, for it melts as Wax; and being melted, poure it into the Mould of Box-wood as aforesaid, it will be a red stone clear and transparent as Crystal, red as a Ruby, then make projection therewith, and set the other half again to multiply.

Then take in Gods Name twenty parts of *Saturn*, which is prepared by Solution and Coagulation, till it leave no more *Fæces* behind, as hath been said at the beginning. Dissolve these twenty parts of *Saturn*, dissolve by itself in a Glass with distilled Vinegar; likewise dissolve the powder of your Stone alone by it self in a Glass with distilled Vinegar, and when both are dissolved into clear water, poure both the Solutions together into a great Glass, set it into a Bath, a head on, and a Receiver to it, distil the Vinegar from it in the boiling Bath, till the Matter be dry, then let it cool of itself, put it into a Glass, lay a polish'd Glass over the mouth of the Glass, and set it into a Furnace in a Cuple with sifted Ashes, make a fire under it like to the Suns heat in *March*, till the powder be perfect white, which is quickly done.

Then augment your fire from one degree to another, till the Matter become yellower and yellower, to a perfect yellow; then increase it yet stronger, from one degree to another, till it be redder and redder, to a perfect redness; then poure your water upon the red powder with the red Oil, or with the water of Paradise, or with the clear sharp Vinegar, or call it how you will, doing in all points as hath been taught, till the red powder flux like Wax upon a Silver Plate, without fuming, penetrating it as Oil doth dry Leather, that it become good Gold within and without; then render thanks unto God, be obedient to him for his Gifts and Graces.

You may again take one half out of the Glass, and make projection, setting the other half in again, as hath been taught, so may you work all your Life-time, for the poor, and perform other duties to Gods Glory, and Salvation of your Soul, as I have said before; enough to the wise.

Projection upon Metal.

Know, my Child, how and in what manner you must use this Stone, which makes projection upon *Mercury*, and all imperfect Metals and Bodies of *Mars, Jupiter* and *Venus*, whereof make Plates glowing hot, whereon straw the Stone, and lay Coals on for a season, that the Stone may penetrate, but the Stones must be made quick with Gold, and *Jupiter* also, which is very laborious, as is taught in the projection. But you must project upon *Saturn* or *Luna*, which need not be made quick, only flux them, and cast one part upon a thousand parts, it will be a Medicine, cast one part of these thousand parts upon ten parts, it will be the best Gold that ever was seen on earth.

Its Use in Physick.

This Stone cures all Leprous people, Plague, and all Diseases which may reign upon Earth, or befal Mankind; this is the true *Aurum potabile*, and the true Quintessence which the Ancients sought; this is what thing whereof the whole Troop of Philosophers speak so wondrously, using all possible skill to conceal its Name and Operation, as aforesaid.

Take of this Stone the quantity of a Wheat-corn, lay it in a little good Wine in a small Glass, half full, or a quarter full, make the Wine warm, the Stone will melt like Butter, and the Wine will be red as Bloud, and very sweet in your mouth, as ever you tasted; for to speak comparatively, it is so sweet in taste that Honey and Sugar may be compared as Gall to it; give this unto the Patient to drink, lay him in Bed, but lay not too many cloaths upon him, the Stone hastens forthwith to the heart, expelling thence all ill humors, thence dilating it self through all the Arteries and Veins of the whole Body, rousing up all humours, the party will sweat, for the Stone opens all the pores of the Body, and drives forth all humours thereby, so that the Patient will seem to have been in the Water, yet will this sweating not make him sicker, for the Stone expels only what is adverse to Nature, preserving what is consonant unto it in its being, therefore the Patient is not sicker or weaker; but the more he sweats the stronger and lustier will he be, the Veins will be lighter, and the Sweat continues till all evil Humours be driven out of the Body, and then it ceases.

The next day you shall take of it the quantity of a Wheat-corn, in warm Wine again, you will go to stool immediately, and that will not cease so long as you have any thing in your Body which is contrary to Nature, and the more Stools the Patient hath, the stronger and lighter at heart will he be; for the Stone drives nothing forth but what is adverse and prejudicial to Nature.

The third day give the like quantity in warm Wine, as aforesaid; it will so fortifie the Veins and Heart, that the party will not think himself to be a Man, but rather a Spirit, all his Members will be so light and lively, & if the party will take the like quantity of a Wheat-Corn every day for the space of nine days, I tell you, his Body will be as spiritual as if he had been nine days in the terrestrial Paradise, eating every day of the Fruit, making him fair, lusty, and young; therefore use this Stone weekly, the quantity of a Wheat-Corn with warm Wine, so shall you live in health unto the last hour of the time appointed for you by God.

What say you, my Child, is not this the true *Aurum potabile*, and the true Quintessence, and the thing which we seek? It is a spiritual thing, a Gift which God bestows upon his Friends, therefore, my Child, do not undertake this Divine Work, if you find your self in deadly Sins, or that your intent be otherwise than to Gods Glory, and to perform those things which I taught you before.

I tell you truly, you may see the Work, or begin it, but I am certain you shall never accomplish it, nor see the Stone, God will order it so, it will break, fall, or some one Disaster or other will happen, that you shall never see the Stone, or accomplish it. Therefore if you find yourself otherwise, do not begin the work, for I know assuredly, you will lose your Labour; wherefore deceive not yourself. Enough to the wise.

Of Natural and Supernatural Things, and Others

Its Use in External Diseases.

My Child, there are some people who have external Distempers on their Bodies, as Fistulaes, Cancers, Wolf, or evil Biles, or Holes, be they what or how they will, &c. give him the weight of one Wheat-Corn to drink in warm Wine two days, as is taught before, the whole body within and without shall be freed from all which is adverse to Nature, and you shall deal with the open Sores thus;

Take a Drachm of the Stone, seeth it in a pottle of Wine in a Glass, the space of two or three *Pater-nosters*, that the Stone may melt, the Wine will be as red as Bloud, therewith wash the Sores morning and evening, laying a thin Plate of Lead over, in a short time, as in ten or twelve days the Sores will be whole; and give him every day the quantity of a Wheat-Corn, in warm wine till he be well. If they be Fistulaes or other concave Holes, that you cannot come at them, to wash them, then take a Silver Syringe, and inject of that wine into them, it will heal home, as aforesaid.

And if one had a pound of the rankest Poison in the world in his Body, and immediately drink a Drachme thereof in warm Wine, the poison shall forthwith evacuate by siege, together with all the evil Humors in his Body.

My Child, here ends the most noble and precious Work which is in the Vegetable Book; on whomsoever God bestows this Stone, needs no other thing, in this World, therefore keep it as close and well as you can, to Gods Glory, who grant that we may walk in his obedience, *Amen*.

God is blessed in his works.

FINIS.

Lightning Source UK Ltd.
Milton Keynes UK
UKHW010628040121
376386UK00001B/217